Hmm ...

George Dlouhy

'I want to know God's thoughts … the rest are details.'
(Albert Einstein)

© George Dlouhy 2019
www.dlouhy.info
National Library of Australia Cataloguing-in-Publication Entry
Hmm ...   ISBN 9781697759891
4. Revised Edition - December 2022

Hmm ...

George Dlouhy

*Dedicated to all
Who reason well.*

Contents

Foreword	1
1. Science Today	3
2. Modeling Our World	7
4. Our Mind	20
7. Human Body	33
8. Laws of Nature and Hidden Constants	37
9. How Our World Was Created	41
10. Living Model	45
11. Light Is Not Mysterious	48
12. The Concept of *Observed light*	53
13. The Light and Us	60
14. Some Uncorrected Errors	63
15. Something to Think About	70
Book's Summary	78

APPENDICES

Appendix 1: Numbers and Our Model	83
Appendix 2: The *Doppler Effect*	88
Appendix 3: The Michelson–Morley Experiment	91
Appendix 4: Lorentz's Transformation	102
Appendix 5: More on Light	110
Appendix 6: The Expanding Universe	118
Appendix 7: Changing the Time	125
The Predicament Song	129

Foreword

My best friend is a person who will give me a book I have not read.
(Abraham Lincoln)

Our world is complex, misunderstood and despite thousands of years searching, we are not even close to comprehending what our world really is. The very first tangible knowledge came during the 6th century BC, most probably from Pythagoras, when he declared that the Earth is spherical. Since then our knowledge has been remarkably enhanced, and yet we still *'know nothing'*, as Socrates, another giant thinker, would surely claim even today. We do not know who and what we are, how we were created, and what the meaning of our life is. Despite so many existing hypotheses and religious teachings, we still did not find the universal, comprehensible, and correct answer based at least on logical reasoning.

The religions are just a faith, not supported by corporeal facts. Isaac Newton's proclamation that the Bible is a code, given to us by God, is unfortunately also only a product of his beliefs and without real proof. It could be argued that his studies in this field were the first, collaterally introducing into the existing scene the role of science This new science challenged the century old practice of just believing, but in an essence, it was still based on faith. The predominant hypothesis is now Darwin's theory of evolution, and it is widely accepted as the only explanation of our existence. Religions, Darwin and all others, offer explanations of our existence, and all claim to be true and yet, some are mutually exclusive. Galileo Galilei's words, *'two truths cannot contradict one another'*, present us with a complicated dilemma. This short quotation does not allow us to accept all explanations of our existence as correct, and therefore we are back where we started: *We simply don't know!*

Evidently, we still have to keep looking. By selecting modeling as a tool, which is basically a comparison between our real world and a simplified and easy to understand model, our task should be made more simple.

There is nothing new about modeling, given that it was already put to practice when humans first invented toys.

For children, toys mimic reality and possess many characteristics, taken from the real world. They help to understand complexities of our world on simplified models, on simple replicas of some real, complex objects and situations.

The modeling in this book follows a similar scenario, analyzing real objects and situations by using abstract and real models. With the help of modeling, this book ambitiously takes on revealing or at least simplifying some of the mysterious aspects of our existence on this planet.

Even this simple approach already derives tangible guidelines, which lead us to a conclusion that our world is nothing more than a sophisticated computer simulation, similar to a game being played in another world. As computers and programming started many decades ago, I am well aware that I am not the first to suggest this comparison. There must be many of us puzzled by such a close resemblance between our world and the virtual, digital world. Toward the end of the twentieth century, when object-oriented programming was introduced, it became an even more tempting comparison. This new concept of modeling objects, existing in our world, into the digital world of computers is revealing and tempting to draw parallels between these two worlds.

For these reasons alone, the privilege to search for the solution to our existence should be now delegated to creative computer modeling.

In this book I have used a simple model of our world, easy to understand and with transparent foundations. The core values of our model correspond to core values of our world, and functionality associated with objects of our model and our world is similar.

As I slowly progressed through some basic, widely accepted science doctrines, I started to discover not just misleading facts, but even badly formulated theories.

These particular topics require more thorough explanation, and therefore are described in detail in the appendixes. Yet, to understand the basics of this book, these appendices are not necessary.

This book is destined for readers categorized by *Galileo* as those *'who reason well'*. Sound reasoning is the desired prerequisite for progress, since it changes our activity from simple learning to investigative studying.

I am certain there are many of us who share the same values and have already discovered what I am about to reveal. The facts presented in this book are basic, comprehensible and formulate my understanding of our presence in this world.

1. Science Today

*To be humane, we must ever be ready to pronounce
that wise, ingenious and modest statement
'I do not know'. (Galileo Galilei)*

In early stages of our life we simply believe what we were told. We had many questions, and we usually received many answers, which we accepted without a slight hesitation. As we progressed through life, we gathered more life's experiences, and we questioned less and less. Majority of people would consider that stage as the end of their learning curve, and with the acquired knowledge they would happily live till the end of their lives.

Fortunately, many start reasoning and questioning the correctness of some of those learnt answers. Some will find them to be correct, and some otherwise. Some will be content with incorrect answers, but luckily for mankind, some will be not. Although they will strife toward the true knowledge, to their disappointment they find that not all questions could have their true answers. At that point, only very few would find the necessary courage to say the simple, 'I do not know'.

Usually, the members of the professional science fraternity shy away from this simple statement, since they consider it as a devaluation of their knowledge and status.

I remember only one person, Professor Martin from University of Tasmania, who was brave enough to admit that there are unanswered questions in his field of knowledge, for which he does not know the answers.

The prevailing situation in today' science is to keep what was already accepted, even if accepted wrongly and avoid any prestige-costly corrections. The following extract found on the Internet demonstrates this situation very aptly:

' ... *(The situation is) no different from the times when people went against the idea the Earth was round. It's not about what is true. It's about what the educational and governing authorities say they want you to believe and say is true. As a result, almost all professors and scientists are too afraid of being ostracized from their communities and face losing their jobs to speak out against the preposterous "science"'*.

Galileo would today probably again insist '*Nonetheless, it moves!*', and although centuries old, this quote precisely depicts the very same situation, which prevailed in his world, and now prevails in our current world.

Even the methods of suppressing the truth are the same, as they used to be: *'the repeated lie eventually becomes the truth,'* and *'if the presented arguments cannot be disapproved, attack the bearer.'*

Many scientists, in their desperate attempt to justify erroneous theories, blindly promoted from hypotheses without tangible proofs, resort to classifying anybody thinking outside 'their square' as 'cranks', and label any opposing arguments as 'pseudoscience'.

Martin Gardner, well known capacity on pseudoscience, in his books defined the characteristic of pseudo scientists:[1]

1. They consider themselves geniuses.

2. They regard other scientists as ignorant blockheads.

3. They believe themselves unjustly persecuted and discriminated against because recognised scientific societies refuse to let them lecture and peer-reviewed journals ignore their research papers or assign them to 'enemies' to review them.

4. Instead of sidestepping mainstream science, they have a strong compulsion to focus on the great scientists and nest established theories. For example, according to the laws of science perpetual motion machines cannot be built. A pseudo scientist builds one.

5. They often write complex jargon, in many cases using terms and phrases they themselves have coined. Even on the subject of the shape of the Earth, you might find it difficult to win a debate with a pseudo scientist who argues that Earth is flat.

These five definitions are not a true evaluation, but a simple character assassination. According to these criteria, Galileo would be even today branded as a pseudo scientist.

The general trend is to accuse anybody not agreeing with a given doctrine to be a conspirator and not believing is branded as a conspiracy. Logically, it is the other way around. Not believing is not conspiring, but substituting a lie for the truth definitely is.

How desperately today's science clings to inaccurate doctrines could be demonstrated on an example of Einstein's special theory of relativity, which is based on incorrect, and incomplete understanding of Michelson-Morley experiment and *Lorentz's* calculations.[2] It is hard to understand how a simple, and even erroneous substitution of the time delay for the change of the rate of time flow, could be overlooked and considered as a valid base for modern physics.

---

[1] *'Fads and Fallacies in he Name of Science'* 1957, from Wikipedia

[2] Chapter 14. Some Uncorrected Errors

How the unexpected result of *Michelson–Morley*[1] experiment could be used to justify elevating time to a status of eternal mystery?

There are many other examples of mishandling of available facts by the current scientific establishment. The phenomenon of '*crop circles*', for example, is one of the simplest. The circles are geometric patterns, appearing mostly in the farmed cereal crop fields, and are made of slashed plants. They started to appear during the 1960s, and during the 1980s were most prolific. At that time, these patterns have developed from simple circles to more complex fractal patterns, generated on a computer.

There were many explanations of that phenomenon, some purely practical, like that the circles were manually made by some enthusiasts, with spare energy and time to waste.

However, to replicate overnight the shapes of fractals would be very difficult, especially when some of the fields covered an area of many hectares.

Some scientists suggested the circles are the result of some unusual weather pattern, some other explanations involve aliens, and others believe the circles are the result of testing some cosmic weapons.

Despite all uncounted articles, debates and publications, no real cause was revealed yet. Closest would be the testing of some remote-controlled weaponry, but the argument against it is the fact that the crop was not damaged, but it was merely flattened.

Since fractals were generated on the computer, and were popular and widely circulated on the Internet for anybody to use, we can exclude aliens and freak weather from the potential culprits.

The very clear and simple explanation is that the circles were man-made, obviously, and were the results of testing some remotely controlled vehicles.

These vehicles were equipped with some slashing attachment, so their path could be distinct and easily recognizable. This is a very understandable and logical explanation and I cannot believe that scientists are evidently ignoring it.

Some of them were surely involved in testing of those remotely operated vehicles and yet, they did not come forward with the truth. If this testing is classified as a military secret, then I wonder what the public is allowed to know, if anything at all.

[1] Appendix 3

Now, if you still do not believe this explanation, then go and ask the farmers, who own those fields. I never met a farmer, who would not know what is happening on his fields.

Considering the absence of substantiated explanations of controversial topics, I felt that even the educational institutions are not the right place to start looking for the right answers. The students there are required to simply learn and replicate whatever knowledge is currently accepted as a norm. The recent developments, like restricting any alternative views, in some universities exacerbated my beliefs and I started to look into all different books written on this subject.

In the very first book on this subject I have read, I naively expected to find some tangible, logical answers. At the end of that book, and actually all other subsequent books, I was disappointed, since the answers were just not there.

As if that was not enough, some reasoning and conclusions in some books were obviously incorrect and I lost faith in my sources. Believing in Socratic *'the unexamined life is not worth living'*, I started my own individual quest.

2. Modeling Our World
'Imagination is more important than knowledge.'
(Albert Einstein)

Comparing the world of digital computing to our world reveals many similarities. We could create a very simple digital model of a tiny part of our world, and although it could be simple and abstract, it could still adequately resemble the world we live in. This model, described and used throughout this book, is obviously still open to any additions, corrections and modifications.

It could be, for example, a simple computer game in which one could excel without knowing anything about the inner workings of the computer. We could create a computer game, mimicking our existing world, and we would use it when attempting to solve some of our world's complex issues. For our game we don't even need a computer and we could create such a model entirely in our mind only.

Just imagine a screen with a picture of a figure in the middle, we would call *Tom*. We could then add some background, and some other persons and objects.

COMPONENTS OF OUR MODEL

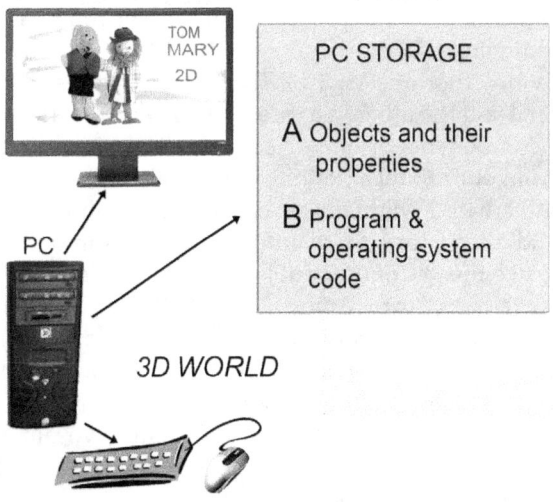

FIGURE 2.1 Existence of our model in 3D world.
(2D = two-dimensional, 3D = three-dimensional.)

Because the computer screen is flat, our model is only two-dimensional.

That brings into our modeling a welcomed simplification and in our three-dimensional world it offers many examples already existing all around us.

The selected model of our world consists of a computer with data storage, computer's display and lines of executable code. The computer exists in our three-dimensional world and images displayed on the screen are two-dimensional representations of some processes, being executed in the computer's memory.

Every object created in our programming model consists of its code, stored in the computer's storage, and it could be also represented by its image, displayed on the screen. Lines of object's code are reachable only from our three-dimensional world and objects in our two-dimensional model have no means of inspecting or changing them.

Let's assume that the program starts automatically and it will initially display two images, *Tom* and *Mary*. For our model to adequately represent our world, *Tom* and *Mary* should be able to mimic human's behavior, part of which is the ability to observe other objects' images on the screen.

Regarding *Tom's* actions as being the result of execution of functions stored in *Tom's* object's code, passing their results to *Tom's* brain means all the programming in our model is done in our three-dimensional world.

It is obvious that one part of *Tom's* existence resides in two-dimensional and the other in a three-dimensional world. Yet, both parts are components of one object only. The three-dimensional part forms in our model a definite entity, called *Tom's mind*.

We could create *Tom's mind* as a part of the computer's storage. *Tom* would then store there information acquired during his existence, like images of other objects, for example *Mary's* image, together with *Mary's* properties.

When *Tom* looks at some objects displayed on the screen, he becomes the *observer* in his two-dimensional world. He 'sees' other objects and creates their images in his brain. These images are one-dimensional, since *Tom's* world is only two-dimensional. Therefore, *Tom* registers only one-dimensional line of observed two-dimensional *Mary's* image.

One-dimensional images, for example the number line, in the two-dimensional world of a computer screen could be displayed as a line, which is actually two-dimensional.

Even if very thin, it still has its width equivalent to the width of the smallest luminous cell on the screen.

In our three-dimensional world, such a line could be represented by a line drawn on a blackboard. The line on the blackboard is just a layer of deposited chalk. It has its length, width and even its height, and therefore is three-dimensional.

When *Tom* looks at *Mary*, her one-dimensional image is created by *Tom's* brain. *Tom's* brain, being a part of *Tom's* image, has limited capability to store observed images. Therefore, they are stored in *Tom's* mind, created on the computer running our model, and existing in our three-dimensional world.

In our world, one-dimensional images cannot exist; therefore, the program would have to store them in the computer's storage as lines of code. Whenever there will be a need to recall these images, our model's program would create them from the lines of code and use them.

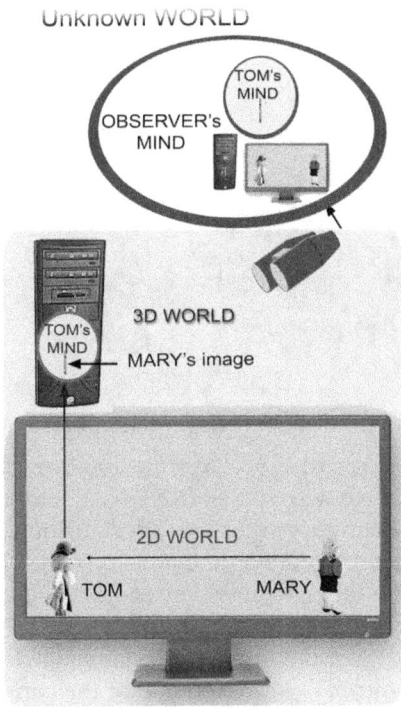

FIGURE 2.2 Tom observes two-dimensional Mary's image and forms one-dimensional image of her.

Tom and *Mary* displayed in our model could be programmed to react to certain events, for example, when seeing *Mary* for the first time, *Tom* would say, 'hello *Mary*,' and *Mary* would reply, 'hello *Tom*.'

At this stage, *Tom* and *Mary* are nothing more than two-dimensional robots and on some given impulse they simply follow instructions, already written into the code of our model's program.

When a two-dimensional image is displayed on the screen of our model's computer, it is obvious that this image, same as a one-dimensional image, cannot exist in our three-dimensional world. After all, it is nothing more than rays of light, processed by the *observer*'s eyes and brain to form an image.

For *Tom* to exist in our world, *Tom*'s image on the screen has to be observed, i.e. there has to be an *observer* in our three-dimensional world, looking at *Tom*'s displayed image.

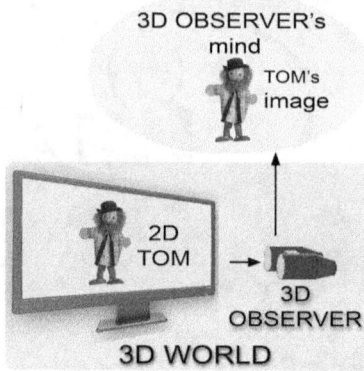

FIGURE 2.3 Tom's image in observer's mind.

By looking at *Tom*, the 3D *observer* in its mind could also associate *Tom*'s image with some of *Tom*'s properties, for example, *Tom* being male. Such properties are called attributes and together with *Tom*'s image and defined functions form the object *Tom*.

From this follows that *Tom* and *Mary* could not reach <u>each other's mind,</u> since their minds exist in computer's storage, located in our three-dimensional world, which is inaccessible to them.

An *observer* in our 3D world, looking at the computer display, will form from the observed image an object in its mind. It will automatically add to the observed image its attributes, and store them all as one, whole *Tom* object.

Tom's object could be programmed, for example, to multiply two infinite numbers √2 x √2, which proves that *Tom's* mind is infinite. Since *Tom's* mind is part of the *observer*'s mind that implies the *observer*'s mind must be also infinite.

Given that the *observer*'s mind is infinite, contrary to the limits applied to the computer screen, objects in the *observer*'s mind could also be infinite.

In our model, only the *observer* in our three-dimensional world could access *Tom's* mind, containing *Mary's* one-dimensional image. At this stage we should emphasize that this described scenario applies also to more than one three-dimensional *observer* of our model's screen.

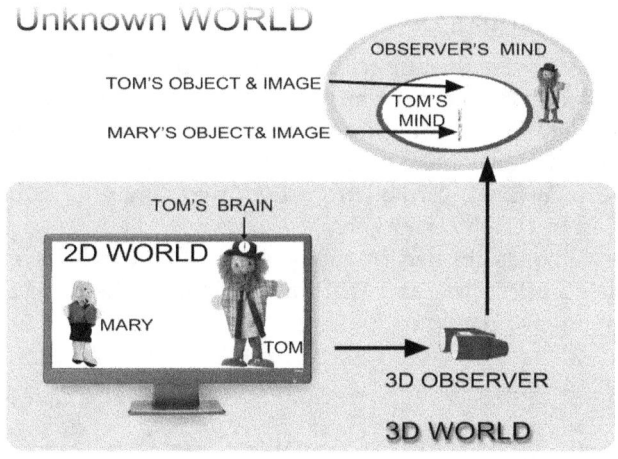

FIGURE 2.4 Tom is looking at Mary, their images exist in a two-dimensional world, i.e. they are displayed on the screen. Observer exists in a three-dimensional world.

When the displayed image is observed by more than one *observer*, the same objects exist in the minds of all *observers*. *Tom's* images become identical parallel objects or even whole parallel universes.

Any visible changes, applied to the observed object, will instantaneously propagate to all objects and all such objects or universes are the same.

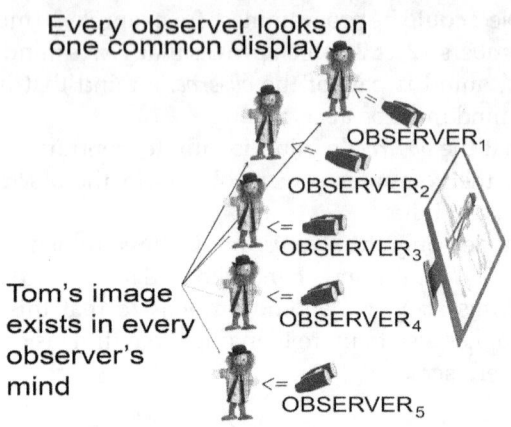

FIGURE 2.5 More observers of the same image.

So far, we were dealing with *observers* in our three-dimensional world, observing a flat screen of our two-dimensional model.

However, they could also observe three-dimensional objects in our three-dimensional world. Images of these objects will then be three-dimensional, formed from two, two-dimensional images, supplied by the *observer's* eyes.

These images formed in the *observer*'s brain will have some associated attributes, and will form objects in the 3D *observer*'s mind, similar to *observers* looking at our model.

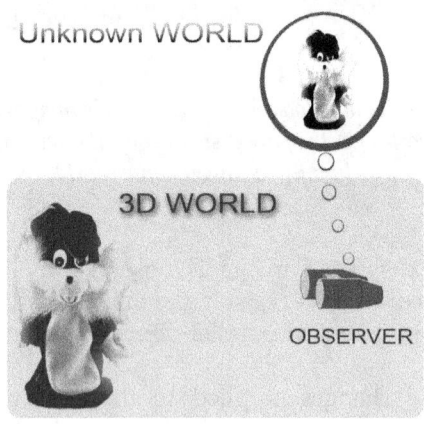

FIGURE 2.6 Observer sees three-dimensional object and forms an object in its mind.

We might conclude that being 3D *observers* in our world, our mind, same as *Tom's* mind in our model, is also a product of a program, running on a computer. *Tom's* mind is created by a computer running in our world and our mind is created by a computer running in a different world than ours. It is not our three-dimensional world, but some other, unknown world.

In our model we could see that there are two distinctive parts of every object: Written code with attributes and the object's displayed image.

Tom's object's code and attributes form *Tom's* mind, which could be compared to a human's soul, and *Tom's* image to a human's body. This could apply generally, and it could be then correctly assumed that all objects have their soul and body.

Implementing this analogy to our world, our body and our soul form an incredibly and meticulously well designed object.

The world of our model is observed by us, existing in our world. Our world, in turn, is observed by *observers* existing in a different world than ours.

Then the paradigm of our model applies also to our world:
- When our body is observed, it becomes an object in the mind of an *observer*, existing in a different world.
- Since not all the objects in our model must have their image displayed on a computer's screen, we could assume that the same applies to our world. We could feel there are some objects in our world that we cannot see, but we could feel or derive their existence.

For example, we could not see the force of gravity, only its effects. Another example are hidden constants, like Plank's constant, existing in our universe. Many of us, including me, also believe that there is God, and claim they could feel God's presence. In our model God could be the programmer, who created the program, and/or one of the *observers*, observing and controlling displayed objects or both.

Programmers and *observers* of our model live in our three-dimensional world, and are not part of our model's world. Similarly, programmers and *observers* of our world live in different world and are not part of our world.

To clarify the proposed existence of a different world, we should consider the example in figure 2.7. Three-dimensional ball exists in our three-dimensional world, and on the screen of our two-dimensional computer model is represented by a two-dimensional image.

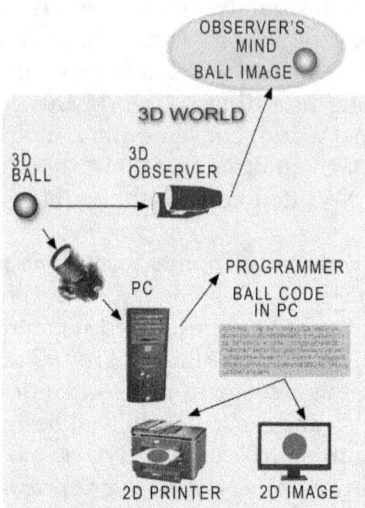

FIGURE 2.7 Transformation of a physical object into a computer's memory and human mind.

When we observe the ball, a pseudo three-dimensional image is formed in our mind from two, two-dimensional images of this ball, sent to the brain by our eyes. When we use a camera connected to the computer, it is up to the computer's program to form a two-dimensional image of this ball on the computer's screen or printer.

For us, living in a three-dimensional world, the code of the two-dimensional ball, stored on a computer, is accessible and therefore it is possible for us to find out what attributes and functions the ball possesses.

That does not apply to the code of the ball's image, stored in our mind. We do not have such an access to that different world of our mind, and therefore we do not know anything about the code there.

The similar situation is when comparing objects in our mind to objects in our model, created in the mind of a 2D *observer*. The code of objects' in our model is stored in our three-dimensional world and for them it is not just inaccessible, but also incomprehensible.

It is also obvious that *Tom* could see one-dimensional images only. To see two-dimensional images, he would have to move to a three-dimensional world, and become three-dimensional.

Applying this analogy to our three-dimensional world, objects existing around us, including humans, do not have access to that different word.

We know nothing about the code used to program our world, and all our knowledge is based on derived, calculated or observed facts; the already mentioned hidden constants is a good example. We can only assume what is real and what are our beliefs.

The progression from lower dimensions to higher dimensions in space, with corresponding changes in dimensions of images of the observed objects, is demonstrated in the figure *2.8*.

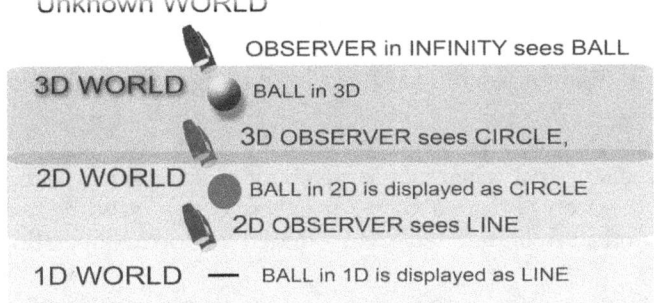

FIGURE 2.8 Object ball exists in the 3D world.

The object ball could exist in the 1D world only as a line, in the 2D world as a circle, and as a ball it could exist only in 3D world. Progressing from the world of higher dimensions to lower dimensions, the image of the observed ball will lose one extra dimension.

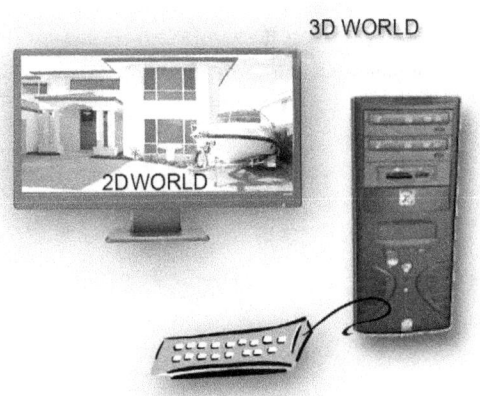

FIGURE 2.9 3D world on a computer's monitor.

In our model we could program objects in such a way that their visual image on the display corresponds to our expectations. For example, a house could be a photograph of existing three-dimensional house.

When we observe a picture of the house displayed on the screen, as in figure 2.9, we see a two-dimensional image of an object, existing in a three-dimensional world. That represents a view of our model, the way a three-dimensional *observer* sees it. It could be said that the *observer* looks at displayed, two-dimensional, images of our model from 'above'. Furthermore, the *observer* sees the whole display, i.e. sees everything everywhere.

Logically considered, the same applies to the *observer* of our world. Then the religious belief that 'God is looking at us "from above" is omnipresent and knows everything,' has some plausible merits.

Some people, who lived through the 'near death experience' also believe that when they were about to leave this world, they saw themselves from above. For example, they saw themselves lying on an operating table, surrounded by doctors operating on them.

We are now in a position to believe that our selected, two-dimensional model is a simplified simulation of our world. In this model could exist objects like humans, animals, mountains, rivers, wind, noise, etc.

We could be satisfied that in this stage of our modeling, our model also corresponds to proven specifics of quantum mechanics, which is a science in physics, dealing with miniature dimensions in the subatomic world of atoms, electrons and similarly small particles. It is a complex science and American physicist Richard Feynman once characterized it:

'I think I can safely say that nobody understands quantum mechanics.'

However, with our model we have already achieved to explain three basic postulates of quantum mechanics:

- *When an object in our three-dimensional world is observed by the observer in a different world, this object then exists in the observer's mind.*

- *If an object in our three-dimensional world is observed by more than one such observer, this object is created in all observers' minds and exists in many identical instances. This is in agreement with the principle of quantum mechanics, called quantum superposition.*

- *Should anything change on the observed object, this change is instantly propagated to all images of that object in all observers' minds. This is in agreement with the principle of quantum mechanics, called quantum entanglement.*

3. Uncomplicated World
Where the senses fail us, reason must step in.
Galileo Galilei.

It should be now clear that we exist as objects in the *observer*'s mind, consisting of our image in our three-dimensional world, and of our mind in the *observer*'s world. As explained in appendix 1, our mind is infinite, and therefore must exist in *infinity*.

The initial modeling has already explained the seamless transformation of images into objects, and therefore our existence is similar to our model: We live in three-dimensions, but our mind, being infinite, must exist somewhere else. To put it simply, we exist as sophisticated, self-aware and complex three-dimensional images in our three-dimensional world, with our mind existing in the infinite world.

The whole situation could be compared to an advanced two-dimensional computer game, like for example the flight simulator, existing in our three-dimensional world. After starting the flight simulator, we become completely entangled in its scenario, and we forget that besides the world displayed on the simulator's screens, i.e. the simulator's two-dimensional world, there is also our three-dimensional world.

In our model we could delegate *Tom* to represent us by his image on the computer screen, and by his attributes in the computer's storage.

Tom's image is two-dimensional though, it belongs to a two-dimensional world and it is not a part of our world.

Judging this situation from *Tom's* viewpoint, *Tom* feels that he lives in his two-dimensional world of the computer screen. He does not know anything about our world, and he could only speculate about our existence. We are *Tom's observers*, and in our three-dimensional world *Tom* exists only in our mind.

We could apply this model to our world and it would be easy to conclude that we are also a computer simulation, only much more complex.

We have already defined our mind being infinite and not being a part of our three-dimensional world. It exists in *infinity*, what in some religions is called Heaven. There is not any reliable description of this world and there is little for us to find out about it.

By selecting a two-dimensional world of a computer game, our world becomes connected to that world by *Tom's* object. Since the light is the crucial factor in creating images, it is obvious that it also forms a connection between these two worlds of different dimensions. By considering the role of light, we could reveal at least a small piece of this enigmatic world of *infinity*.

One of the attributes of the light is its speed, which is a ratio of distance to time. In three-dimensional space, the light in time period $\underline{t}$ would cover distance $\underline{D}_{Light}$, which is equal to:

$\underline{D}_{Light} = t \cdot c_{Light}$ (*time traveled* × *speed of light*)

In the *infinity* exists only *subjective time*, as later explained in chapter *4*, which is infinite: $t = \infty$. The distance is also infinite: $\underline{D}_{Light} = \infty$,

therefore, the above equation becomes $\infty = \infty \cdot c_{Light}$
and $c_{Light} = \infty / \infty$, which is **undefined**.

(Please note: c_{Light}, i.e. ∞ / ∞ is **not** equal 1, otherwise the following will apply: $\infty / \infty = 1$ and $\infty + \infty = \infty$
Then $\infty / \infty = (\infty + \infty) / \infty = (\infty / \infty) + (\infty / \infty) = 1 + 1 = 2$)

We could conclude that the speed of light is undefined in the *infinity*, and therefore the light cannot exist there.

This is obvious when we consider that we could clearly see images in our mind and yet, there is no need for any light to display them. Definitely, we do not need our eyes to recall any images embedded in our mind.

Since in *infinity* the speed of light is undefined, it is understandable that our world, where the speed of light is defined by the physical structure of the universe, cannot be infinite.

The presence of our three-dimensional image in our three-dimensional world could be only a broadcast of events, happening in the *infinity*. It could be similar to a school blackboard, which depicts results of processes taking place in teachers and pupils' minds.

The very notable difference is that unlike objects on the blackboard, we are equipped with self-awareness, we are mobile, capable of making our own decisions, and we feel the existence of our body in this three-dimensional world.

The ability to sense our surroundings and feel our presence in our world is a part of our code, which is similar to programming in our model. Objects there could be equipped with the capability to sense their surroundings, like for example *Tom* could walk into some obstacle and react to it.

Since we have an infinite mind, we must be a part of an infinite world, i.e. the *infinity*, and we are obviously not just robots. Our image is being observed by the *observer* in *infinity*, and our mind fuses with the *observer*'s mind.

We could even entertain the possibility that we are the actual *observers*, observing 'from above' our own three-dimensional image, existing in a three-dimensional world.

We can conclude:

- *Objects in the two-dimensional world of our model, created by the program, are observed in our three-dimensional world using the light. Images of these objects are formed in the mind of three-dimensional observers, i.e. our mind, which is a part of infinite space called the infinity.*

- *The objects in our three-dimensional world are observed from infinity. Observed objects form images in the mind of observers there, and as a result of that our real existence is in the infinity. Similar to our model, these images are results of computer simulation created by the program, running our world from infinity.*

- *Our three dimensional world is not infinite, as is generally believed. The presence of defined light renders this concept impossible.*

4. Our Mind

'A person starts to live when he can live outside himself.' (Albert Einstein)

We have already established that our mind is infinite. As a consequence of that it cannot exist in our world, but in some new, unknown world. Comparing it to *Tom's* mind in our model, it becomes obvious that *Tom's* mind also does not exist in *Tom's* world, but in our world, known to us, but not to *Tom*. The question then remains, what is this new, unknown world?

Regrettably, in our reasoning we have only one tangible guideline to follow: Since our mind is infinite, it must exist in some infinite space. Then this unknown world must be also infinite, and in this book we already called it the *infinity*, the world already used to generally denote anything without ending. In addition to our mind, this new *infinity* world must also contain all the code running our world.

Initially, we know nothing about our world, but as we progress through our lives, we are constantly learning. We gather various information that enables us to form our opinions and beliefs. I am certain there is nobody who could say that they would not do things differently, should they start from the beginning of their lives again.

During the learning, our knowledge is perpetually accumulating, and the amount of stored information grows.

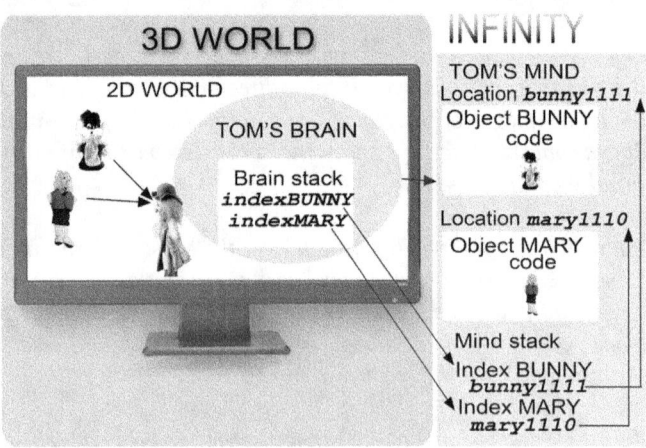

FIGURE 4.1 Tom in our model world observes Mary and Bunny.

In our model we have already defined the computer storage as an object's mind, containing all the information of each individual object. The same analogy applies to us and to our mind. This raises many questions, of which the most important are:
- What is our mind, and what does it represent?
- Does our existence have any purpose?

Tom, being a 2D *observer*, looks at Bunny, and *Tom's* brain forms the object Bunny in *Tom's* mind. Object Bunny consists of Bunny's image, and some code describing Bunny's attributes, like Bunny having big ears, etc.

The same applies to *Mary*, and all other objects *Tom* observes. It is obvious that *Tom's* brain, being only a two-dimensional image on the screen, does not have the capacity to store all that information.

The solution to this problem is to store all objects' information on the computer's hard drive. That requires the model's computer to keep the addresses of stored information, and use them when retrieving the object into the object's memory.

When *Tom* is observed, i.e. he 'lives', his mind is part of the *observer's* mind, and therefore is infinite. It could store an infinite amount of information, requiring an infinite number of addresses. *Tom's* brain does not have such capacity, and therefore these addresses have to be kept in *Tom's* mind.

For *Tom* to access them, he has to keep a track of these stored addresses in his brain. Thus, we would have to create two stacks of addresses, one with limited capacity in *Tom's* brain, which we would call brain stack, and one with infinite capacity, we would call mind stack.

The reason for using addresses only is that they are just numbers, and require much less storage than the whole object's information code. These addresses are usually called indexes or pointers.

When the program is running and *Tom* sees an object, his brain will form the object's image, add to it some attributes, and then pass the whole new object as an information bundle to the program. The program will store this bundle in computer storage, and add its storage's address to the list of addresses on the mind stack, stored in *Tom's* mind. Then the address (of this stored address on the mind stack) is added to *Tom's* brain stack.

For example, when *Tom* sees *Mary* for the first time, his brain will form *Mary's* information bundle, containing her image and attributes, and pass it to the program.

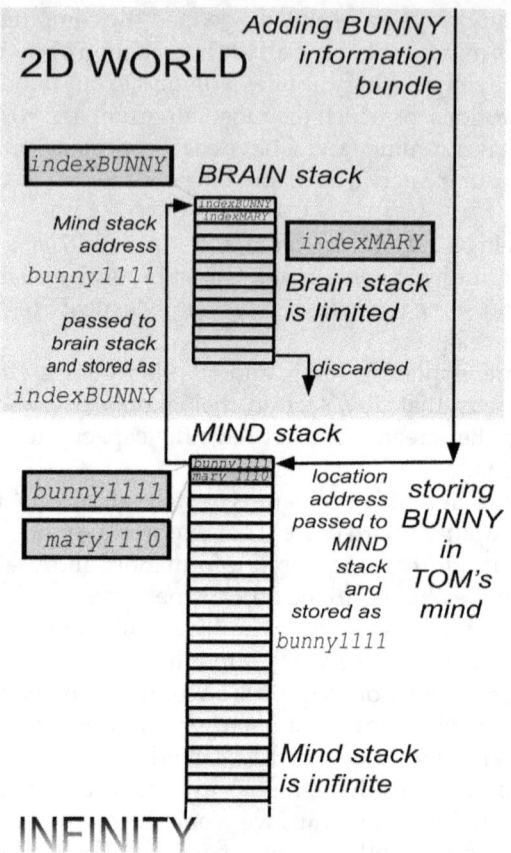

FIGURE 4.2 Processing of newly acquired information bundles. Brain stack resides in the object's image and the mind stack in the object's mind.

The program then stores this bundle on the computer's hard drive, and creates an address `mary1110` on *Tom's* mind stack, pointing to the stored information.

To finish, the program adds an address (of this address) to *Tom's* brain stack. In our example, this address is `indexMARY`, and it points to `mary1110`, the second entry on the mind stack.

When *Tom* looks at Bunny, his brain will form Bunny's information bundle, and pass it to the program. The program then stores this bundle on the computer's hard drive and creates an address `bunny1111` on *Tom's* mind stack, pointing to the stored information. Then the program adds the address (of address) to *Tom's* brain stack, as `indexBUNNY`, pointing to the first entry.

To retrieve *Mary's* information bundle, *Tom's* brain searches the brain stack for `indexMARY` and from the mind stack obtains the address `mary1110`. It passes it to the program, which then proceeds to recover the stored information bundle from this address on the hard drive.

Applied to our world, it is obvious that the amount of information stored in our mind is limitless. What is not limitless is our brain, containing the brain stack.

It is believed that human brain's can reach a maximum capacity of 2.5 million of gigabytes or whatever.

Although this a large number, after considering the amount of daily encounters with new objects, and the length of human's life, this capacity could not be adequate to permanently store all the indexes on the brain stack.

Therefore, most of the brain's capacity is taken by the brain stack, which leaves only a smaller part of the brain to be used to perform all other tasks. This is the functional part of the brain, and takes care of the running human body, decision making, feelings, behavior constraint, consciousness, etc.

With some allowances for the accuracy of observed functioning of our brain, we could assume that the brain stack is organized on the principle FIFO, i.e. 'first in, first out'[1], which is the common principle used in computer database management.

The index of last information acquired is stored on the top of the brain stack, and all other, already stored indexes, are shifted down by one position. If the capacity of the brain stack is reached, the index of first information stored, being at the bottom of the brain stack, is deleted.

This is the underlying basic flow of events, supported by observed proceedings during our lives. Already known in ancient Rome was 'repetitio est mater studiorum' i.e. repetition is the mother of learning. Any particular index in the brain stack of a healthy person could be refreshed by acquiring or just recalling the same information again and again. This information bundle and its indexes will be newly added, and the bundle will be available for retrieval as currently acquired information.

Storing and retrieving information from our mind is affected by many factors, for example the age, illness, brain capability, etc. During severe loss of brain cells, the size of the brain stack reduces drastically and our brain could not store any more indexes.

[1]The other principle is LIFO, *'last in, first out'*.

That means we register all objects around us, store them in our mind, but we cannot recall them through our brain stack. This is despite the fact that once the information is registered by our mind it stays there, together with its address on the mind stack. Our mind is not subjected to the conditions of our body, but our brain stack is. Occasionally, some damaged brain cells also cause their corresponding addresses on the brain stack to be lost, and then it is impossible to retrieve the information to which the address was pointing to.

Under special conditions, i.e. during deep sleep, narcosis, dreams, hypnosis, etc., it is possible for the brain to access the information bundle stored in the mind through the mind stack directly and not via the brain stack.[1]

This is a very concise description of the inner workings of our brain and mind, and in reality it is much more complex. The storing of indexes on the brain stack is definitely affected by the frequency of usage of any particular information bundle.

Is is also very probable that any frequently used information is pushed into the 'permanent' part of the brain stack, while seldom used information is kept in the 'volatile' part. That would explain why we usually remember some unique, often recalled information about what we did many years ago, but do not remember what we did yesterday.

The everlasting existence of our mind implies that our existence in this world cannot be meaningless. What we have learned during our life and whatever information we acquired, will be retained in the code, which has created us. That translates to our existence in *infinity* being infinite.

Closely related to our mind is also the time phenomenon. The time in our world represents only the past, and there is not a tangible present, and definitely not a future in our world. Whatever is happening, the moment when it happens becomes already the past. We should call this time the *universal time*, which was already defined by Isaac Newton many centuries ago, and is known as *Newtonian time*. The real present in our world is so infinitely small, that it could only exist in the *infinity*, where we live. This *universal time* is probably the most important attribute of our world. It is not flexible, and it proceeds with its unchangeable rate of flow.[2] Our lives therefore start and end at some given instance of progressing *universal time*.

[1] See chapter 5. Hypnotism
[2] See chapter 14. Some Uncorrected Errors

The same situation applies to our model. For example, *Tom's* image could start when the computer is switched on and ends when is switched off.

In addition to the *universal time*, there is another form of time, what we shall call *experienced time* or *subjective time*. This is the time experienced by each object in our model separately.

We could program into object's code a clock, ticking sometimes slower and sometimes faster. It would be different for each object and it would run independently from the *universal time*. The rate of *subjective time* flow will change with circumstances and the state of mind of individual objects.

The same time model applies also to our three-dimensional world. This will become clear, should we consider that our mind contains past events, which we could sort chronologically and we could recall them again and again. In our mind we could go back in time without changing time on the clock, ticking on the wall.

This means that in our mind also exists the *subjective time*, experienced differently by every one of us. It is only our mind, where we could recall past events and we could also fantasize about the future.

The existence of such *subjective time* could be demonstrated, for example when patients on operating table under narcosis have their senses 'switched off', and the flow of their *subjective time* stops. Operation could take hours and yet, when patients wake up it seems to them that the operation did not take any time at all. Falling unconscious and waking up from anesthesia seems to them to be instantaneous. It is obvious that during the operation patients' *subjective time* stopped, while *universal* time, applicable to the whole universe, kept going at its normal rate.

In the *infinity*, the *universal time* consists only in its present form, which is infinite and has no bearing on our *subjective* time. That signifies that the *infinity* is eternal.

We can now answer the initial questions:
- *What is our mind, and what it represents?*

Our mind is permanently stored on the computer running our world and while we are being observed, our mind is also a subset of our *observer's* mind, existing in *infinity*. That means that our mind exists in the *infinity* and therefore is infinite.

During our lives in our world, our brain registers all experiences and information and transports them from our temporary, three-dimensional world into our mind in permanent *infinity*. It connects our existence in this world with our existence in the *infinity*.

- *Does our existence have any purpose?*

The existence of our mind is the main factor why we exist. While we live, we learn and enhance our knowledge, which is permanently accumulating in our mind. Our mind is the foundation of our existence in the *infinity*.

We could use our model and ask *Tom* what he knows about his existence in his world? He simply would not know. With the exception of his world, created by our model, he would not know about the existence of anything else.

He would be in the same position as we are, being asked what we know about our existence?

However, should we be asked about *Tom's* existence in our model, we could give many qualified answers. It is our world, after all, where the program of our model exists and runs, and we know all the information and reasons for creating it.

That implies that it is not we, who could come up with any concrete answers to the purpose of our existence of our mind in *infinity*.

However, now we are at least in a position to make a qualified guess, a firm, logically plausible answer, based on what we scrutinized so far.

We can conclude:

- *Our brain exists in our world, it is a part of our body, and its capacity is limited. The acquired information during our life is stored in our mind, and not in our brain. Our brain only tracks where each piece of individual information is stored in our mind. The biggest part of the brain is assigned to the storage of these indexes and only the smaller part is used to fulfill the functional role of the brain, needed by our body.*

- *Our mind is unlimited, and exists in a world other than ours, called infinity. In our world, the only factor limiting the amount of information accessible in our mind is the capacity of our brain, i.e. how many pointers to stored information our brain can keep.*

- *Universal time is valid for the entire universe. It has a constant rate of flow, which cannot be changed* [1]. *It could be measured differently in different reference frames. In Earth's reference frame, the rate of flow of universal time is measured by the atomic clock.*

[1]See chapter 14. Some Uncorrected Errors

- *Universal time is an absolute entity and in our world it consists only of the past. It flows in the direction from past to the apparent present and future. The present part of the universal time exists only in the infinity and is eternal.*

- *The subjective time is the object's experienced time. It is specific for each living individual object, it exists in the object's mind, is infinite and could be moved back and forward. Its rate of flow could change for each individual object.*

- *The study of acquired information forms in our brain opinions and conclusions and enables us to replace the wrong information in our mind with correct one. This process is called learning and is the inevitable answer to the question of the purpose of our existence in this world. This process of learning enables us to become a better person, which is the ultimate purpose of our lives.*

- *Our existence in infinity and its purpose is known only to our creator and for us it will remain shrouded in mystery.*

5. The Hypnotism
All that we are is the result of what we have thought. (Buddha)

Despite the long history of hypnotism, this ancient phenomenon already mentioned in the Bible, is even now still enshrouded in a mystery. Christ was able to cure sick people and the ancient civilizations used hypnotism to restore health and suppress pain.

There exist many tangible records of such practice and one of the most remarkable devotees of hypnotism was doctor Franz Anton Mesmer[1], who lived in Vienna during eighteen century. The English language even invented words *'mesmerize, mesmerism'*, derived from his name.

His most known achievement happened in 1777, when Maria Theresa Paradise, infant prodigy and blind pianist, recovered her sight after his treatment.

At that time, she was for ten years already under the care of Dr. Von Stoerk, best oculist in Europe, but parently without any improvements.

---

[1] 'HYPNOTISM AND THE POWER WITHIN' by Dr. S. J. Van Pelt

The very recent and revealing demonstration of hypnosis I have encountered was one of the episodes in excellent documentary series Mythbusters, created in 2007.

The team decided to investigate should the hypnosis be considered a myth or not. Three of the team, Grant, Kari and Tory, were firstly tested for suitability and then they were unknowingly exposed to a live episode, involving actors: Two actors delivered a package, had a brief, and even heated conversation and then they left.

Each member of the team then received a questionnaire, relating to that played episode, and was asked to answer all questions as much as they remembered.

During the second part of the experiment, the team members were hypnotized, and then received new questionnaire. While hypnotized, they answered the same questions again.

In the first questionnaire were only a few noticeable observations, but in the second questionnaire, done under hypnosis, the answers were describing in more details what actually happened. For example, they remembered the displayed name on men's name tags and the description of the tattoo one of the men had on his neck. The conclusion of their experiment was that hypnosis is for real and it is not a myth.

When the team members were hypnotized, they seem to fall into induced, different state of mind, often referred to as subconsciousness. They were still able to answer any questions, but their spiritual presence in this world was somehow altered.

To find an explanation to this experiment, we could try our model, and let *Tom* hypnotize *Mary*. Her brain becomes partly blocked, and fully responsive is only her mind, residing in our 3D world.

When *Tom* is usually talking to *Mary*, he is communicating through her brain, which in turn communicates with *Mary's* mind. During hypnosis, the *brain stack* in *Mary's* brain is blocked and *Mary's* brain accesses stored information in *Mary's* mind directly through her *mind stack*.

The *mind's stack* is limitless, but the *brain stack* is not. It is also slower to respond and therefore not all the addresses of stored information, ever registered by the *mind stack*, are necessarily registered by the *brain stack*.

That explains why under hypnosis, the team members were able to remember more details from the staged episode.

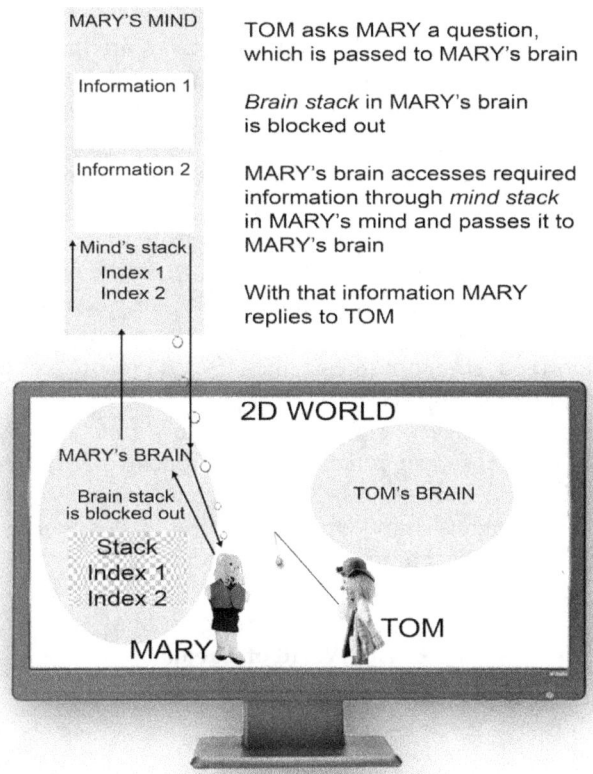

FIGURE 5.1 Tom is hypnotizing Mary.

The results obtained in Mythbusters experiment suggest that the hypnotized team members' brain communicates directly with their mind stack, same as does *Mary's* brain in our model.

Similar to hypnosis is anesthesia, used to induce the state of low or no responsiveness at all.

The difference between anesthesia and hypnotic stage is the intensity of reduced activity of the brain. The anesthesia induces deep sleep, during which the person does not respond at all. On the other hand, hypnosis is more likely compatible with a less intensive stage of awakening from anesthesia.

At this stage of awakening, there is a mischievous practice to ask such a person some very intimate questions. Usually that person will reveal all, what otherwise is closely guarded information. In a fully awakened state, personal feelings and developed moral constraints will prevent this from happening.

This implies that the personal feelings and constraints are stored in the person's brain. Good examples of them are the feelings of joy, guild, remorse, pain, etc. These feelings and constraints are part of our body, i.e. the body feels pain, joy, noise, movements, guilt, remorse, feelings of what is right or wrong, etc. They are specific to a particular, ever-changing point in *universal time*, and since this time does not exist in *infinity*, where our mind exists, personal feelings and constraints cannot be a part of our mind.

They are relevant only to the brain and when the mind is being accessed directly through the mind stack, these constraints and feelings are not registered there. When the brain is partly switched off under hypnosis or anesthesia, the affected person does not feel any pain, hunger or craving.

There is definitely a separation of bodily functions and the person's mind. We can conclude that hypnosis partly bypasses a person's brain and connects the hypnotized persons directly to their mind. Persons' mind does not reside in this world, like a person's brain does, but in an infinite world.

6. The World of Dreams
'Reality is merely an illusion, albeit a very persistent one.'
(Albert Einstein)

Establishing that our mind exists in an infinite world creates an inevitable question: How does our mind exist in such a world? The answer to that could be found by answering another question: How do we exist in our dreams?

Our dreams we experience during our sleep represent a mysterious combination of imagination, knowledge, and events. They are like an interactive movie, where we are the main actors. We 'see' our dreams through our partially subdued brain, and sometimes we even intensively feel all associated emotions. The time involved in our dreams is our *subjective time*, with an uncertain rate of time flow. Same as during hypnosis, our dreams do not access brain stack. At most, they only create emotions and feelings in our brain.

Objects' images and attributes in our dreams are accessed through the mind stack. That explains why, when we wake up from a dream, soon after we forget what the entire dream was about. The brain's direct connection to the mind stack becomes broken, and is redirected to the brain stack again.

Many could argue that dreams simply reflect some unintentional processes, mistakenly being carried out by our brain during our sleep.

Contrary to that, I believe there is some reality involved in our dreams. For example, many people are having dreams, during which they speak one or more languages they learnt. That knowledge is real and not a pure fantasy, created by our brain.

There is no plausible explanation of what the dreams are and there are many myths surrounding them.

Some people believe that the future could be predicted by analyzing dreams. If they are correct that means the future is already established and is defined somewhere.

One of my friends, whose integrity I do not doubt, is adamant that in some cases he can predict the future. He supports his beliefs by examples when he correctly did so.

Yet, for him to predict someone's future, this future has to be already predetermined, which would have to apply to all other existing, interrelated objects. They would have to follow some predefined scenario, and they all would have to be mutually synchronized. To implement it would be only possible, if we are dealing with a very limited number of objects.

For example, for *Tom* to meet *Mary* in a restaurant, and *Mary* to meet *Tom*, they both would have to be programmed to do so. When we add more objects, actions of every one of them would have to be synchronized with actions of all other objects.

Such a scenario will be not just impossible to program, but all the decisions of all objects, right from the beginning, would have to be also predefined. That would put our model on par with some passive game or movie, for example. The decision making will be taken away from individual objects and our model will not represent our world any more.

What is then the explanation of my friend's belief in his prediction of the future? He obviously believes in his ability, without being aware of it, and by some lucky coincidence he was able to predict some events. Since the future does not exist, trying to predict it from one's feelings or dreams is pointless.

The dreams we experience during our sleep are often described as strange and unreal, and often are depicting some strange places, not known to us.

In these dreams we also experience everything more intensely than we do in our real world and we often wake up completely devastated by the way our dreams proceeded.

For example, I am able to recall very vividly one of my repeating dreams, where I live in a city, so intimately known to me. Even after waking up, I could see its high, sunlit walls, reaching down to the wide, azure sea bay, with a bright, stony island in the middle. During my life so far, I never saw such a city and I never met people I encountered in my dreams. In my dreams I could effortlessly levitate and I was always surprised that nobody there considered it as being anything special.

In one of my dreams I also saw myself in a mirror and I could not recognize myself. I looked quite apart from my mirror image in this three-dimensional world.

That Mediterranean city, I dreamt of so many times, is every time the same. It is embedded and stored in my mind and its existence could be a result of some forgotten, previous experiences. After all, our mind is infinite and does not vanish with our death.

While dreaming, we usually do not feel our body, and actually we do not feel anything, except some emotions, which are the products of our brain. Through our brain our body feels emotions we go through and reacts accordingly. For example, we cry or laugh in our dreams and simultaneously, and completely without our consent, our body does the same.

Sleepwalking would be another example of our body communicating directly with our mind. After waking up, the persons do not remember at all they were walking, and do not remember anything they have encountered during the walk. Their brain was subdued and the brain stack did not register any events.

We have already established that our mind is infinite, exists in *infinity* and is affected only by our individual *subjective time*. During sleep, the brain stack is disabled and the brain communicates with the mind through the mind stack. As a consequence of that, we could access all information ever stored in our mind.

Similarly to hypnosis, our dreams are only a small and temporary window into the infinite world.

7. Human Body
The ultimate beauty in this world is the beauty of a woman.

Our presence in this world is only a part of our existence. Our body represents our mind in this three-dimensional world, and therefore is also three-dimensional. We live in this three-dimensional world, but our mind is not a part of this world, since it exists in *infinity* only.

The bony skeleton supports muscles, all our organs and enables us to move. Three main parts of our body are the brain, coordinating all activities, the lungs and the heart, supplying oxygen, and stomach and intestines, supplying fuel, necessary to make the body operational.

Lungs will provide oxygen, stomach and intestines provide the sources of energy and the heart and blood will deliver it all to the muscles.

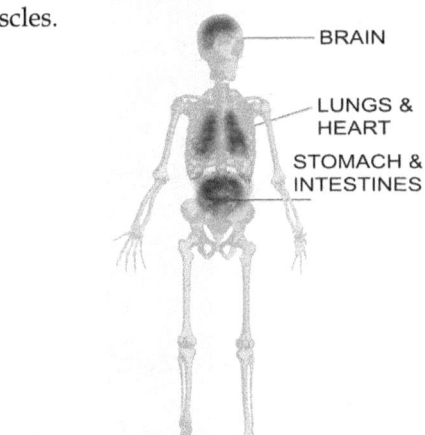

FIGURE 7.1 Basic components of the human body.

There, the processed sources of energy will be oxidized, i.e. burnt, which produces energy, enabling the functioning of muscles.

This functionality of the human body could be broadly compared to an engine burning fuel, producing energy and turning it into a work.

Development and coordination of these processes requires a deep knowledge of chemical reactions, which could be, to some extent, programmed into our model. Should we believe that such knowledge has evolved, then this possibility should be clearly demonstrated by our model.

Yet, in our model nothing evolves by itself. The code of the model or some of its pseudo, self-programming functions, could be changed only in our world, where the code is stored.

Whatever we need in our model has to be programmed in advance before we can use it. The same analogy also applies to our world.

Like many computer programs running in our world, even the program running our world has its flaws. One such quite remarkable flaw is generally known as *'An Incendiary Enigma'*. There are more than hundred reported incidents, where the human body was inexplicably consumed by a fire. The victims burst into flames, which destroyed only the torso and head, but not the limbs, which is contrary to the normal burning process.[1]

Although this phenomenon is still considered by many as a mystery, there is a plausible explanation.

We could use our model and investigate the possibility that *Tom*, i.e. his image on the computer screen, would disappear in such a violent manner.

Firstly, we would have to program into *Tom's* image a program sequence, happening in our bodies, which will emulate oxidization of hydrocarbons. This should not be a problem, since such a code could be created. Oxidizing is nothing more than burning, as we know it, only in this case is controlled, and only the appropriate amount of hydrocarbons needed is oxidized.

That requires an inclusion of controlling code, which will decide the beginning and duration of such burning. This code must rely on sensors, inbuilt in the body, which will force decisions to be made by the controlling program.

Without going into any details, it is already obvious that any malfunctioning of the program would be mainly due to the faulty sensors. If the sensors, reporting to end the burning, malfunction, then the burning would not stop and will continue to oxidize any organic material in its vicinity. That will presumably consume the whole body.

We could encounter the same situation when lighting a campfire in the forest. Sitting around the fire, we control the burning to an acceptable level. Should we let the fire burn without any restrictions, soon the forest will start burning too.

The last remaining enigma to be explained is why the limbs are not burned as well as the body and head?

[1] Source 'Mystery of the Human Body', Time-Life Books

To answer that, we should firstly consider where the oxidation of hydrocarbons in our body is usually happening?

Even if we are motionless, our brain still functions, heart is still beating, we still breathe, and our digestive system, with all the supporting organs, is still functioning.

That points exactly to these parts of our body, where in a case of sensor's malfunction, the uncontrolled burning usually happens.

Such burning requires oxygen, which is initially delivered by the blood, and in exposed parts could be also taken from the atmosphere. While the fire is spreading inside the body, the only oxygen supply enabling the fire to continue is the blood. When the organs and body are damaged to such an extent that there is no blood flow into the limbs, and the fire has not reached the atmosphere yet, the burning ceases.

That would be similar to placing a lit candle into a jar and putting the lid on. Such a candle will be burning for a short period only until all oxygen in the jar is consumed by the flame.

By using our model, this explanation of spontaneous burning is plausible. There are hardly any programs or hardware in our world, which would not have some unforeseen faults, associated with corresponding malfunctioning.

These faults could lay dormant and might be never activated, unless some weird combination of events or some unpredictable malfunctioning of the hardware will evoke them.

Given that we are also programmed, there is no reason not to believe that the above described situation could eventuate in our world. After all, there are already many people living with visible flaws in their creation.

We refer to our body as 'ours', but we do not fully control it. What we control are the movements allowed by the design of our body. We control what we eat and drink, and what we observe. To some extent, we control when to breathe, when to eat and drink, when to discharge our body waste, when to sleep, etc. Yet, there is a lot that we do not control.

Our brain is bombarded with many different chemicals, produced by our body's different organs, which are inducing our feelings. We feel joy, sadness, but also we feel angst and pain. We fear death, what serves as a deterrence against ending our life prematurely, and we feel pain, preventing us from willfully damaging our body. Then we have some essential feelings, of being hungry and thirsty, tired, hot and cold, healthy, and the most rewarding feeling from them all, the feeling of being well.

Our body even tells us what it needs. The obvious feeling of being thirsty or hungry is complemented by craving for some certain food. I still remember how eagerly I ate large grapefruit, skin and all, after waking up from a three-day long bout of sickness in Rangoon.

Sometimes I could eat chilies and enjoy them, yet sometimes I cannot, and that despite them being harvested from the same little garden in our backyard.

Our body has many ways of telling us what it needs, and it does not have to be by taste only. For example, after excessive drinking of alcohol, the stomach content becomes alkaline. Since our stomach content supposed to be slightly acidic, an alkaline reaction will trigger an alarm somewhere in our brain. At that instance, the brain will usually develop a taste for pickled cucumbers, and if these are not coming soon, it will force the stomach to empty itself of its undesirable content.

It is easy to remedy this situation with a few pills of vitamin C, which, being acidic, will restore the stomach's acidity level back to normal.

There are some decisions, which our body makes without any regards to our wishes, and without even letting us know.

We could start this series with the simple decision when to fall asleep, and finish with the decision when to die. The inescapable aging leads to this procedure automatically, and there are circumstances when our body has to make this decision based on our health and condition. When our body is seriously damaged, it evokes firstly a shock to the system, which suppresses the pain, and then the body conducts damage assessment. If the damage is not considered immediately lethal, the shock changes into a pain, and we still live. In case of lethal damage, our body simply decides to die. Our body represents us, i.e. our mind, in this three-dimensional world, and thanks to our body we realize ourselves in this world. Yet, we are nothing more than guests in our body, sitting at some limited control hub, listing to our body's desires and deciding what to do next.

The Human body is in its essence very similar to a mechanical robot. It has its frame, control hub, represented by the brain, it has its sensors, consisting of eyes, ears, nose and skin, etc., and it has its own motors, consisting of many muscles. Most importantly, it has us to drive it. Considering all this, one has to admire this marvelous piece of engineering.

8. Laws of Nature and Hidden Constants

*'Not everything that counts can be counted,
and not everything that can be counted counts.'
(Albert Einstein)*

In our two-dimensional model we have already defined that there could exist many different objects. Some have their image displayed on the computer screen and we could see them, and some are hidden, although they exist. It depends purely on how we have programmed our model.

Some objects could be complex and some plain. We would expect the code of a chunk of flour dough to be simpler than the code of a piece of rubber. After stretching, the rubber returns usually to its original shape, and therefore contains some specific 'memory' and appropriate functions, which after stretching would restore the initial properties of that piece. The dough does not have that.

In our computer model the code of such a two-dimensional image of a piece of rubber, displayed on the screen, could be accessed, and we could inspect it and change it. That does not apply to a piece of rubber in our world. Its code is not in our world, since it is stored in the *infinity*, and therefore is not accessible to us.

As in our model, in our world also many different objects exist, some we could see and some not. Although we could not see them, we could feel or calculate them. These hidden objects could be just a number, for example the gravitational constant $G = 6.67 \times 10^{-11}$ N m^2/kg^2. This is an example of just one of the hidden constants existing in our world. Besides these constants, there exist also laws governing our world, for example laws of thermodynamics.

The question is how such constants and laws were defined?

It was already discovered by *Willem de Sitter*[1] that the light is progressing relatively to the medium with its constant speed. Why exactly this speed? And how is it possible that our eyes rely on the value of this constant? Why were they designed in such a way that exactly this speed of incoming light ensures the light to be registered and processed by the eyes?

We could question further and ask why the laws of thermodynamics exist?

Nobody from our world has invented such laws and nobody defined such constants. Their existence was found experimentally, and their values were calculated from observed parameters.

---

[1] More details in appendix 3.

There must be even more laws of nature and more constants, which we have not discovered yet. They rule our world and we could sense their existence.

One such mysterious phenomenon is randomness of events, which could be demonstrated in figure 8.1.

During this experiment, after each draw we add a small bead to the tube, representing the drawn number. For example, when we draw number 5, one bead is dropped into a tube marked '5'. Then the ball is returned into the drum and drawings continue.

If we repeat this process, say 30 000 times, we will discover that the number of beads in the tubes, representing frequencies of drawn individual numbers, is increasing at almost an even rate. It does not happen that one tube will be filled considerably more than any other.

Even if this is not proof that there is a law governing the randomness of these draws, it is an indication that some laws must be governing the functions of this apparatus.

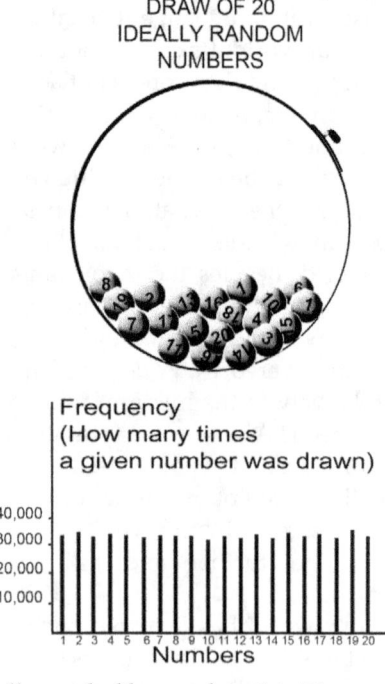

FIGURE 8.1 Balls, marked by numbers 1 to 20, are randomly drawn by ideally random apparatus. (The draws of such apparatus are not influenced by any physical imperfections.)

Another example is casting of a dice, when the number 3 appears on the top after three, successful castings. Would you bet on that same number again? Answering 'no' means that you believe this event keeps its own history somewhere!

Should we look at this scenario logically, with every new cast, the number 3 has the same chance to appear on the top, as all the other numbers. Yet, there is something urging us to believe that after three consecutive casts, always with this number on the top, its chances to appear on the top again are getting smaller.

From these two experiments, we should not be burned at stake for believing that there is a law governing randomness in our world. To validate our belief we can use our programming model and try to program similar mechanisms into its code.

Firstly, we should define randomness: When dice is cast, the numbers on top must not form a repeating sequence and their frequency must keep a broad parity with all the other numbers.

To achieve it, we would with every cast check all the numbers already registered, and make sure that the new number is not a part of an already drawn, repeating sequence. That would require keeping a database, where we store all numbers already accepted, and all sequences formed. After every new cast we would then check this database.

Should we find that so far there is no sequence being formed, then we proceed further and compare all the frequencies of these numbers. For example, we could set the allowable percentage of possible difference from the average to 2%. If the frequency of the newly cast number exceeds that, it will be rejected and a new cast would start. When both criteria are satisfied, the new number will be accepted.

At the start, this process will be fairly simple, but with the increasing number of casts, it would become more and more intricate. For example, for a typical lotto draw, the number of all different combinations to be checked is 3,838,380!

Six-numbered group ($k = 6$) out of 40 ($n = 40$) without repetition:

$$40! / (6! \times (40-6)!) = 3,838,380$$

With an increasing number of draws, the program would have to check an astronomical number of combinations and our random function will soon collapse. Our two-dimensional model in our three-dimensional world is not capable of ensuring truthful randomness, but it proves that for a small number of casts, the randomness could be programmed into our model.

Definitely, in our world we can see the results of random processes. Typical example would be the process of forming a human's face. Obviously, the creator of our world used a code that we cannot replicate.

What we can replicate in our program are hidden constants, and some basic laws. In our model we call them the 'global constants and functions,' and as such, they are rigid and valid for the whole program and all created objects.

The values of *global constants* cannot be changed and they are established at the start of the program and discontinue its existence only when the program ends. This comparison with our model explains the existence of hidden constants in our world - they were simply set by the physical structure of the universe, and are unchangeable parts of our world. We have discovered many such constants, but there is still possibility that there are many more, waiting to be discovered.

An excellent example of one discovered natural law is the law of triangles, discovered by Pythagoras[1], and obviously, the randomness of processes in our world is an example of yet undiscovered law.

We could conclude that the laws of nature and hidden constants were set, when the universe was created. From our world we cannot change these laws and values of these constants. Some are also built into the code of our body like for example, the constant speed of light in vacuum,. Human eyes could see only the light, approaching them with this constant speed[2].

This represents two independent entities, the light, and the eyes, both accommodating the same defined constant c. The time of their creation differs vastly, and it is obvious that the eyes were designed only after the constant c was already established. For evolution to achieve this, it would mean that during their creation, the eyes would be able to acquire the exact value for speed of light in the air and include it in their design. That makes the concept of the eyes' creation difficult to attribute to the evolution.

[1] More details in appendix 1.

[2] More details in chapter 13. The Light And Us.

9. How Our World Was Created
'The eternal mystery of the world is its comprehensibility.'
(Albert Einstein)

We are now in the stage where we have clarified basic concepts of our existence in this world. With the help of our modeling we have already established that we are nothing more than objects, formed in the mind of an *observer*, living outside our world. We have also explained the basic structure and behavior of our body and mind, and explained the mysterious existence of unseen structures in our world. Another interesting topic to discuss is how our world was created.

There were, and still are, many answers to this question and many are still considered to be valid. Different groups of people have their own answers and needles to say, they are all based on beliefs only.

With the exception of modern science, all these answers are based on beliefs in God, and are not based on any empirical evidence. Science does not believe in God, but with its explanation of our existence does not fare better than the religions do.

Since the program, which created our world resides outside our world, we have no means to access it and therefore our explanation would not offer any concrete proof either. All that we could offer in this book are logical arguments only.

Obviously, the task of creating our world was enormous and it is beyond our imagination to comprehend it. All we can do is to simplify this process on our two-dimensional computer model, which is an abstract and only two-dimensional model of our world. We could go through the world's creation, step by step, and see how this process would compare with reality.

The very first, logical task would be to create in our model two-dimensional universe, and we should be able to access any part of it.

We set our computer screen to be just a tiny window into this universe, which could be moved in any direction and any distance. Furthermore, one of the window's property will be a zoom, use to display this universe in more detail. All that is possible to create in our program and is acceptable to use it in our model.

The next task is to create some content, i.e. the Sun, planets, etc. We enjoy the comfort of instantaneous creation due to the fact that our model is abstract, and the world's image in our mind could be easily evoked.

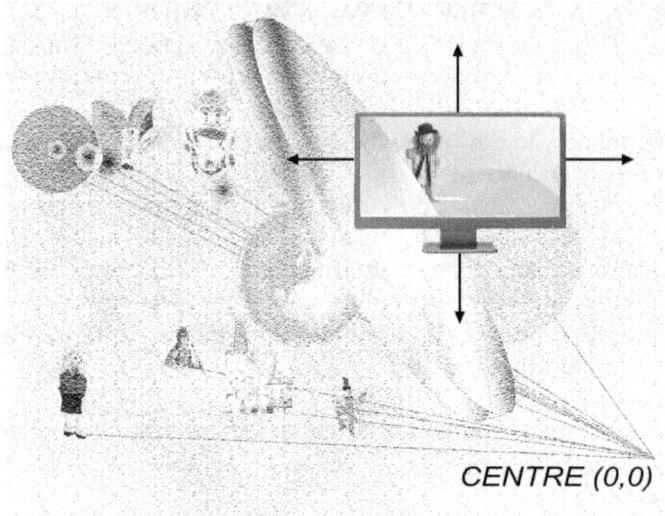

FIGURE 9.1 *Our model of the universe.*
The computer screen represents a window, which could be moved in any direction.

All we have to ascertain is that the desired creation of our two-dimensional model is feasible.

One notable feature of this process so far, is the fact that nothing has evolved and everything was created.

The next task is the creation of life in our model. *Tom* lives while the object *Tom* exists in the computer's memory and *Tom's* image is being observed. *Tom's* code enables him to perform different tasks, which could be considered as a sign that he lives.

If the evolution theory is correct, and life progresses from low forms to higher forms, then the program would have to be capable of creating some more code, and add it to the lines of code already written. Can our computer do it?

Our model's program can certainly learn. It can store acquired information and then make decisions based on the nature of that information.

The only problem is that such decision-making code has to be already programmed, before that decision is made. The only chance for any new code to be automatically added by the program to the existing code, would be *'self-programming'*, i.e. program's ability to modify itself. Let's see what that means for our model.

For program to write a line of executable code, it would have to know the programming language, and it would also have to know, where such a line of code should be inserted into the program. Both of these tasks are evidently impossible for our program to perform.

That leaves us with the only possibility: Our model's program has to be fully created, and nothing in it could have evolved. Without some additional programming, done by the programmer in our three-dimensional world, there could not be any changes implemented in the program's code of our model.

To apply this scenario to our world, it is obvious that we cannot have any input into the structure of the code, running our three-dimensional world. We do not know the code, used to create our world, we do not know how and where to send such code-modifying instructions, and most importantly, we have no access to our code. Without a divine intervention, objects in our world could not change their code's structure.

What objects could change are the object's attributes. They are not part of the object's code, but are stored in object's mind, which is specific to every individual object, and is accessible by the objects and their *observer*s.

Every newly created object is equipped with a different set of attributes, i.e. some of us are tall, have different color of hair, skin, eyes, etc. During our life, some of these attributes are found to be inappropriate to the circumstances, and therefore are changed. For example, not enough of nutrition will change the size attribute. A typical example of this are undernourished fish. Their size is proportional to the amount of food they can find.

Some attributes are passed by the parents onto the new generation, as for example the brain capacity, face, and body shape, prevailing body colors, etc.

From observations of our world is obvious that for the creation of a new individual, the aim is to select the best attributes, suited to a given environment, and combine them in the mind of the newly born baby. That should produce better individuals than their parents. This cannot be considered as the evolution process, because the underlying code was not changed; only the properties were. Fish are still fish and for many thousand of years, the *Homo sapiens* has not changed into any other, more advanced form.

That process should be correctly classified as adaptation to living condition and the selection of positive attributes during reproduction should be considered as selection of species.

At this stage of creation, the emphasis is on new species of minute dimensions, like bacteria and viruses. There are always new strains of viruses and bacteria being discovered, and there are some species continuously disappearing. It is similar to the trend in the current hi-tech research and development. The miniaturization is an inevitable outcome of progressing knowledge and technology.

So far, all that supports the creation and not the evolution. Yet, there are still many arguments put forward by supporters of the evolution hypothesis, which are supposed to be persuasive and legitimate. For example, the most publicized similarities between the skeleton of a bird and a fish served as the basis for beliefs that birds evolved from fishes.

The more plausible explanation is offered by creation, though. The design called 'blueprint' is an engineering and programming term, and it is used as a foundation for development of new products and new programming code. The structure of such a design is reused in new projects, and only some parts are modified.

That offers a simple explanation of similarities between fishes and birds. Such a 'blueprint' was used to create fishes, later was modified to produce birds, and even later, some more advanced designs were used to produce birds capable of flying.

Similarly, the first humans' designs were also primitive, and only in later stages were they replaced by the current human model. Scrutinizing the defined concept of evolution of humans even more, it is obvious that since the emergence of current, modern humans, no evolution has taken place.

Like the introduction of new species, the selection of species is also continually happening. The main difference between these two concepts is that creation of new species requires changes in their code, which is not accessible from our world. Natural selection is done only by changing attributes of already defined and created species and does not require any changes to the object's code.

Darwin's *'Theory of evolution by natural selection'* incorrectly combines both concepts together, despite they being different. It is similar to mixing apples and pears and expecting to finish with a homogenous mix.

Obviously, our world was created outside our three dimensions, and we have no means of changing the code of its creation.

This logically derived statement is in sharp contrast with the evolution hypothesis, but to some extent it is in agreement with religious teaching. Yet, contrary to the religions, our model reveals that the creation was not as simple, as is believed by the religions.

Our model also proves that there must exist another world, we already described and call *infinity*, where our mind resides, and where we will eventually return.

It should be noted that these conclusions are not just mere beliefs, but they were derived using realistic, logical modeling, which evolution does not offer.

10. Living Model
'Technological progress is like an axe in the hands of a pathological criminal.' (Albert Einstein)

In this book we based the explanation of our existence on the assumption that we are just a sophisticated game, executing in *infinity*, i.e. some unknown, infinite world. When this game is observed by *observers* in *infinity*, we become images with our own mind, being also a part of *observers'* mind.

Since the purpose of this game is to enhance our mind, the game's scenario should inevitably include situations, where we learn what is right and what is wrong. The game should be a never-ending struggle between righteousness and wickedness, between the truth and lies. I believe that there cannot be peace in our world, since this could not be a part of this game's scenario. It would make the game unchallenging, although for us definitely more acceptable.

Instead of guessing what our destiny is, we should better look around in nature and find some already existing examples. If some event already happened, then there is a great probability that it might happen again. We could find some simple model, or at least a model easy to understand, and apply it to our civilization. This time we should choose some model different from the one we used so far, and select one of the living models, found all around us instead.

It should have the same basic parameters as our civilization, namely it has to be some population of living objects, with the ability to reproduce.

Other desirable parameters would be a suitable living space and availability of resources to be consumed.

When limiting ourselves to these requirements, all we have to do is to look in our world for some living model. We give our preferences to a population of objects with a short life span, since we do not wish to wait to see any results.

One ideal example of such civilization of living objects, being present everywhere around us, are microorganisms and the most suitable for our model would be bacteria culture producing alcohol. We can create such a population by simply filling a jar with some fruit and water. Then we add some bacteria culture, available from shops catering for home-brew enthusiasts. This new population has to be kept at room temperature and bacteria will start to consume the sugar in fruit. As with most of living objects, bacteria will also automatically produce some by-products, in this case it will be the carbon dioxide and ethyl alcohol. The first is a gas, and it will escape to the atmosphere, but the alcohol will stay dissolved in the water, where bacteria live. Initially, the bacteria will have no shortage of resources or space and they will uncontrollably multiply.

So far, all that corresponds to what humans are doing on this planet. Only our model proceeds to its end at a much greater pace.

The problem the bacteria have is that the alcohol they produce is poisoning them. When they produce enough alcohol, and the concentration increases to a level they cannot tolerate any more at approximately 17%, they start dying. They were not programmed to comprehend the gravity of their situation and therefore they will proceed toward their own destruction regardless. Soon they all die and their place in this world is taken by some other bacteria cultures. Usually, the bacteria producing vinegar will take over, and we will have a new, slightly different model to observe.

This graph in figure 10.1 only approximately illustrates the life-cycle of a typical bacteria culture, by placing on the timeline the uncontrolled population growth.

The position on the top of the curve represents either depletion of resources or unacceptably high concentration of poisonous by-products. In that stage, bacteria start to die faster than they are reproducing.

Bacteria culture's problem is also one of the problems so typical for our civilization. At the beginning, humans found themselves in a favorable environment with plenty of resources.

POPULATION GROWTH

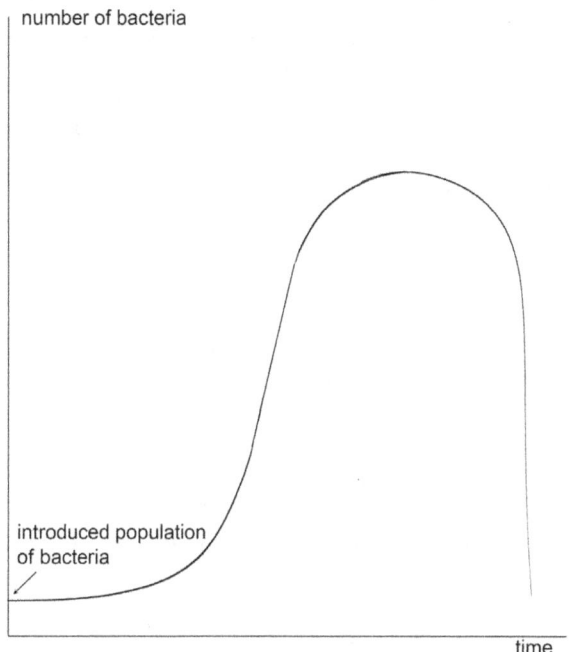

FIGURE 10.1 Representation of a typical life span of bacteria culture.

They started to consume these resources and started to multiply. During this process they produced and released into their environment poisonous by-products and were occupying more and more available space.

Same as bacteria, humans also have limited resources and limited living space. In the history of our civilization, the population density was initially low and epidemics and permanent wars kept the human population at sustainable levels.

The concentration of poisonous by-products was also low and all living objects coexisted in a balance.

Then advances in medical science almost eliminated epidemics and new technologies enabled humans to produce more and more food to support a fast growing population. That brought with it a sharp increase in demand for resources and living space, and a sharp increase in concentration of poisonous by-products.

So far, there is no consensus amongst countries on how to keep population growth on a sustainable level and how to force industries to stop producing poisonous by-products.

This situation corresponds very fittingly to our bacteria model, and should our civilization remain a part of this vicious circle, we would follow bacteria' fate.

Since this pre-defined scenario for bacteria culture was written in the code used to create and run our world, we have no reasons to believe that humans will be an exception and will follow a different path.

11. Light Is Not Mysterious
'There are those who reason well, but they are greatly outnumbered by those who reason badly'.
(Galileo Galilei)

The prerequisite for reasoning well is that any such reasoning must be based on correct facts. Since the reasoning in this book is the fundamental force behind our modeling, we have to first ascertain that the facts derived by the science so far, are valid and unique.

One of the most predominant factors supporting our existence in this world is the light. Unfortunately, today's science considers it as some magical phenomenon, mutually affecting some other factors, like time, mass and energy. However, the light is similar to the already well-understood sound waves.

Material	Approximate Speed (m/sec)	
	Sound	Light
Air	343	300,000,000 [1]
Water	1,490	225,000,000
Rubber	1,500	-----
Glass	5,600	200,000,000
Iron	6,000	-----
Aluminum	6,400	-----
Vacuum	-----	300,000,000

FIGURE 11.1 Maximum measured speed of sound and light in different mediums.

[1] The latest using lasers is 299,792,458 m/sec

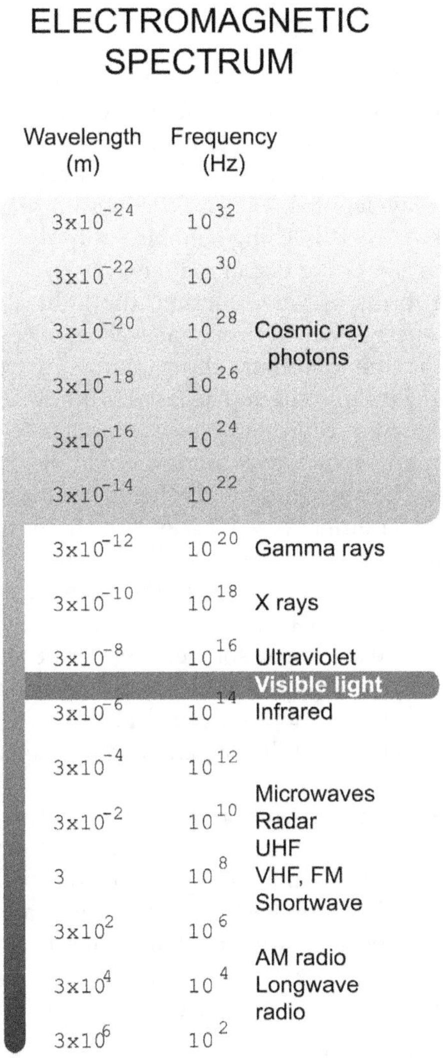

ELECTROMAGNETIC SPECTRUM

Wavelength (m)	Frequency (Hz)	
3×10^{-24}	10^{32}	
3×10^{-22}	10^{30}	
3×10^{-20}	10^{28}	Cosmic ray photons
3×10^{-18}	10^{26}	
3×10^{-16}	10^{24}	
3×10^{-14}	10^{22}	
3×10^{-12}	10^{20}	Gamma rays
3×10^{-10}	10^{18}	X rays
3×10^{-8}	10^{16}	Ultraviolet
		Visible light
3×10^{-6}	10^{14}	Infrared
3×10^{-4}	10^{12}	
3×10^{-2}	10^{10}	Microwaves Radar
3	10^{8}	UHF VHF, FM
3×10^{2}	10^{6}	Shortwave
3×10^{4}	10^{4}	AM radio Longwave radio
3×10^{6}	10^{2}	

FIGURE 11.2 Electromagnetic spectrum.

Both, the sound and light are waves progressing through some medium and have their speed limited by that medium.

The Bible states that the light was created. This is evident once we start looking for some other plausible answers to the basic question: 'How has the light eventuated?'

What we know already is that the visible light is only a small fraction of electromagnetic radiation, which in turn is a part of our physical world. Since the visible light occupies only a small part of that radiation spectrum, we do not see the rest of it.

However, using other sensors than our eyes, we could detect some other wavelengths. We could expose a photographic film to X-ray radiation or by using suitable equipment, for example a radio receiver, we could detect radio waves.

As all the forms of wave motion, the light also starts as an outburst of energy, which propagates as a succession of waves. The bigger is the outburst, the grater is wave's amplitude, representing wave's strength. The faster is the outburst, the higher is wave's frequency, representing wave's pitch.

These properties of wave motion could be clearly seen on ocean waves; the amplitude is the height of the waves, and the frequency is the number of waves, arriving to the shore in a given time interval.

By dropping a grain of sand into the water will create smaller and more frequent waves, than dropping a stone. Another example could be a comparison between a rifle shot and cannon shot. Since the rifle shot produces much lower energy and much faster than a cannon shot, we could hear a rifle shot having a lower noise level (i.e. amplitude) and higher pitch (i.e. frequency) than cannon shot.

Since these waves progress in the same medium, the speed of both waves has to be the same. Therefore, for different frequencies they would have to have the corresponding change in wavelengths.

Both, the sound and light waves share some common properties. In some examples we could treat the sound as light and vice versa.

The wave has its speed defined by its frequency and wavelength:

$$speed = frequency \times wavelength$$

The figure 11.3 depicts an ideal scenario, where the wave is damped by the medium in which it propagates.

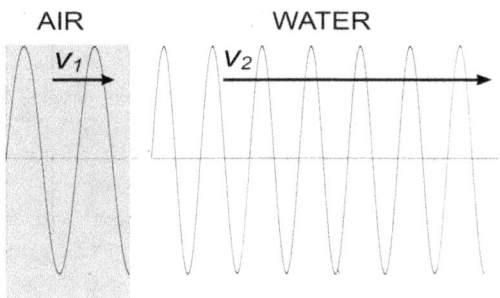

FIGURE 11.3 Sound wave traveling through the air and through the water.

The sound wave starts progressing through the air with the speed $\underline{v}$ = 343 m/sec. A typical frequency of audible sound of a note $\underline{A}$ is $\underline{f}$ = 880 Hz, with corresponding wavelength λ = 0.39 m.

The wave propagates faster through the water, than through the air. For the same fixed period $\underline{t}$ its speed would have to increase, and the multiple of *frequency* **x** *wavelength* would have to be equal to 1,490 m/sec, i.e. the speed of the light in water.

That means the listener in the water will hear the higher pitched sound than in the air.

In the scenario in figure *11.3* we could substitute the light for the sound, glass for the air, and vacuum for the water, and we would have very similar situation for the *propagating light* through different restrictive mediums.

This analogy supports the fact that vacuum restricts the speed of light as any other material. The belief that the maximum speed of light was somehow defined is highly improbable. As we mentioned already, the constants in our world own their existence to the physical nature of our solar system, and are not defined, but derived. They are the result of comparing and analyzing some already existing entities.

We could justifiably treat any medium in respect to *propagating light* as restrictive, including the vacuum, and for time being ignore the dilemma about the presence a substance in vacuum called *aether*.

The maximum speed of propagating waves allowed by the specific medium could be theoretically exceeded by the speed of the wave's source. This is only true, should the restrictive medium allow it.

In this respect, traveling in space is not different to an aircraft, traveling in the air. When such an aircraft exceeds the maximum speed of sound in the air, the sound of the engines will not reach the pilot, sitting in the cabin in front of them. Similar situation applies to a moving *observer* and the light in space.

When the *observer* moves away from the light source with the speed higher than constant speed c of light in vacuum, the light would not reach the *observer*. The *observer* would not see the light coming from that particular light source, but could see any other light, emitted by different light sources.

We could conclude that in the reference frame of a stationary medium, any changes to observed propagating wave are caused by:

1. Medium, limiting the speed of propagating waves. This affects the speed of the wave, its real frequency and wavelength.

2. The speed of the source of light, moving through the restrictive medium. This further affects both, the real frequency and the wavelength of propagating wave, as described by the *Doppler Effect*. [1]

3. *Observer*, moving relatively to the stationary restrictive medium. This affects the observed frequency, and the observed speed, both experienced by the moving *observer*. [2]

When we treat the space and light as any other physical phenomena, i.e. the light as wave and the space as restrictive medium, then we have the light behaving exactly as expected:

- The light propagates through the medium called space with its constant maximum speed c, relative to the medium, and independent of the speed of its source.
- Should the light be considered from the reference frame of a moving *observer*, this will change. The light becomes the *observed light*, and has to be treated differently.

---

[1] More in appendix 2, The *Doppler Effect*
[2] More in chapter 12. The *Observed light*

12. The Concept of Observed Light
'Give me a fixed point and I will move the world' (Archimedes)

So far, we have loosely used the term 'speed' without considering any concrete reference frame. However, the speed is always relative to some selected reference frame. For example, when we ask at what speed the train travels, we usually assume the reference frame being some stationary buildings we just passed by, i.e. we measure the speed of the train relative to the stationary buildings.

In the following example, we first select stationary *Tom* to form a reference frame for all other objects, i.e. the speed of other objects is measured relative to *Tom*. Then we change the reference frame to *ibis*

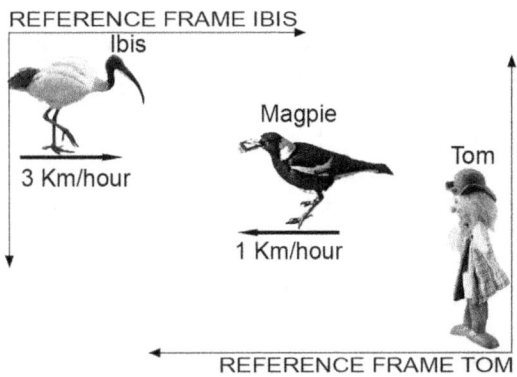

FIGURE 12.1 An example of movement of different objects created in our two-dimensional model.

Reference frame *Tom:*
  Magpie is moving at speed 1 km/h.
  (Moving ahead.)

Reference frame *ibis:*
  Magpie is moving at speed 4 km/h.
  (Moving behind.)

This example demonstrates the general idea behind the relativity of movement. *Magpie* moves at different speeds, depending on the reference frame, to which the movement is related.

There exists a group of people known as geocentrists, who believe that the Earth is the centre of the universe.

Since every movement should be considered in relation to some reference frame, it would be easy to select the Earth as the origin of such a frame and believe that the whole universe is rotating around the Earth. That would be in agreement with the principle of relativity, formulated by Galileo, but the observations of our solar system form a different, generally accepted concept.

To make the situation even more confusing, we have the theory of expanding universe, which is based on observation of the light emitted by distant planets. By comparing the wavelengths of the *observed light*, this theory calculates the speed of such an expansion, and states that the direction of expansion is away from the Earth.[1]

These observations are conducted in the Earth's reference frame and if we change to another reference frame, their results should not be different.

When dealing with the light, especially with its frequency, wavelength and speed, the selection of reference frames is crucial. For example, the term 'constant speed of light in vacuum', generally represented by letter c, is measured only relatively to the vacuum, i.e. one of the mediums in which the light propagates.

For this constant speed of light c, the vacuum is the universal reference frame, which limits the speed of light exactly to this value and any movements of the *observer* or the light source have no effect on the speed of *propagating light*. The term *'propagating light'* is used to describe the light progressing through the medium, without involving an *observer*.

This categorization is necessary, and deserves more comprehensible explanation.

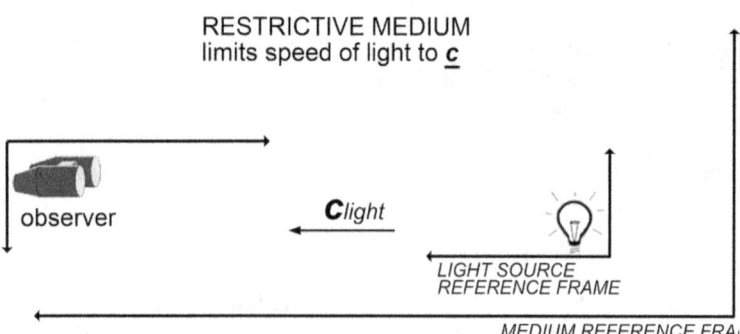

FIGURE 12.2 *The light observed in different reference frames.*

[1] See chapter 16. Something to Think About

The scenario, depicted in figure **12.2**, could be considered from two different, significant reference frames:

1. In *MEDIUM REFERENCE FRAME,* the medium is stationary, the *observer* and the source of light could be also stationary or could move in any direction.

2. In *OBSERVER REFERENCE FRAME*, the *observer* is stationary, and the medium and the light source could be also stationary or move in any direction.

We firstly investigate the *situation 1* as seen from *MEDIUM REFERENCE FRAME,* illustrated in the figure **12.3**, involving a moving light source.

Without the restrictive medium, the light source will produce the wavelength and frequency of emitted light, corresponding to the initial amount of energy of its creation. The speed of light will be unknown.

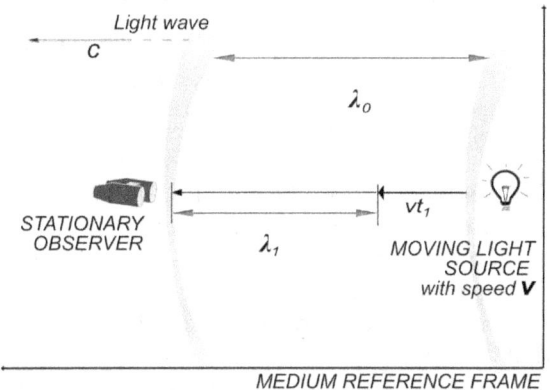

FIGURE 12.3 *The light source is moving through a restrictive medium and towards the stationary observer.*

With vacuum as the restrictive medium, the situation will change, and the speed of light will be its constant speed in vacuum <u>*c*</u>.

When the light source is moving with speed <u>*v*</u>, in time <u>*t_1*</u> will cover the distance <u>*vt_1*</u>. For the same period <u>*t_1*</u>, the light's wavelength λ_0 will be shortened by this distance and becomes λ_1. Thus the *propagating light* will have shifted frequency and wavelength, depending on the speed of the light source.

The speed of the *propagating light* is set by the medium, i.e. the vacuum in this example, and it will remain the same, should the source be moving or not.

A More realistic situation is when both the source of waves and the *observer* are moving independently.

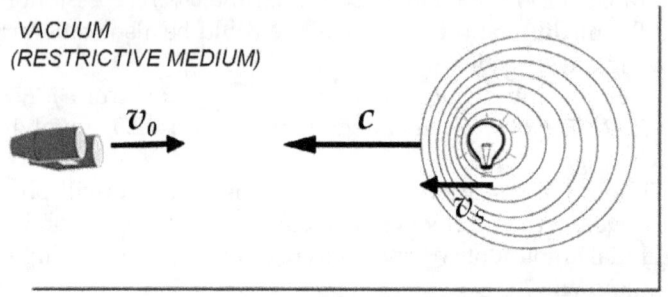

FIGURE 12.4 The light source and observer, both moving relatively to the universal reference frame.

Applied relatively to the universal reference frame, the light source and the *observer* could be both moving. The restrictive medium imposes limit $\underline{c}$ on the speed of light, but theoretically there is no limit to the speed of the light source and *observer*. Feasibility of this scenario would become clear when compared to the sound propagating through the air. The speed of the *observer* and source, relative to the air, could exceed the maximum speed of sound in the air.

So far, we were dealing only with the *propagating light*, which is not the case in this example. The light encountered by the *observer* is the light the *observer* sees. Its speed is measured relatively to the *observer's* reference frame, and is affected by *observer's* movements.

In the following example, we could simplify the situation and consider only the *observer* moving towards a stationary light source.

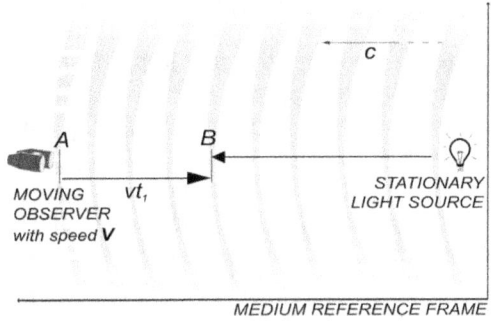

FIGURE 12.5 *The observer is moving toward the stationary source.*

The frequency of emitted light from a stationary light source, measured relatively to the vacuum, is the standard frequency of visible light. When the *observer* moves toward the light source, it encounters more waves per unit time $\underline{t}_1$, which increases the observed frequency, and yet, the wavelength of the light will remain the same.

Since the speed of light equals to

speed of light = frequency x wavelength,

it implies that the speed of light experienced by the moving *observer* will be higher, if the *observer* is moving toward the light source, and lower, moving away from the light source.

To support this statement, consider the following example illustrated by figure ***12.6***.

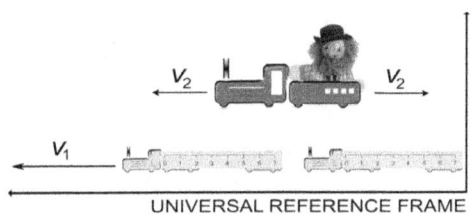

FIGURE 12.6 *Trains carrying rulers, representing the wavelength, are passing Tom's train.*

Tom sees the ruler, representing the light's wavelength, on all wagons in its original length. The length of the ruler does not change, no matter how fast *Tom* or trains travel. Faster the trains travel, more rulers will pass *Tom's* train, and the frequency of passing rulers increases.

It is obvious that the *observer*'s movement is crucial to the observed speed. There is then a fundamental difference between the light propagating through the medium and the same light, being observed by the moving *observer*. To separate these two forms of light, we shall use the terms *propagating light* and *observed light*[1].

The major difference between them is that the frequency and wavelength of the *propagating light* are both affected by the movement of its source. For the *observed light*, its frequency and speed are both additionally affected by the *observer*'s speed.

At the time the *observer* encounters *propagating light*, light's frequency and wavelength could have already been changed by the movements of the light source through the restrictive medium.

By reaching the *observer*, the *propagating light* becomes the *observed light*, and its frequency and speed, relative to the *observer*, will be further changed by the relative movements of the *observer*.

The universe could be filled with *propagating light*, and yet there will be no *observed light*, unless in the universe exist an *observer*.

It is now obvious that each differently moving *observer* will see the same *propagating light* differently. This inevitably leads to a conclusion that the *propagating light* in vacuum moves always with its constant speed $\underline{c}$, but the *observed light* could have different frequency and different speed, which could be slower or faster than light's constant speed in vacuum, $\underline{c}$.

Observed entities could be any propagating waves, observed by *observers*. They do not have to be necessary people, they could be photographic or digital cameras, light-sensitive emulsion, and all relevant types of sensors.

Some *observers* detect not just the light, but also different frequencies and wavelengths of the electromagnetic spectrum; for example X-ray and radar waves, and some other *observers* detect ultrasound, sound waves, etc.

Observers moving relatively to the universal reference frame will detect these propagating waves, and their observed properties will change with the relative speed of the moving *observer*.

---

[1] More on light in appendix *A5*

We can conclude:

- *The **frequency** and **wavelength** of the **propagating light** is affected by the speed of the light source, measured relatively to a restrictive medium. In case of vacuum, the speed of propagating light will be relative to the vacuum reference frame. The constant speed $\underline{c}$ of propagating light in vacuum is not affected by the speed of the light source.*

- *The **observed speed** and **observed frequency**, but not the wavelength of the **observed light**, propagating through the restrictive medium, is affected by the speed of the observer, relative to the restrictive medium.*

- *In our world exist two different types of entities: The **real entities**, existing in the universe, and **observed entities**, created by the interaction between real entities and observers. Observed entities could not exist without observers.*

13. The Light and Us
'In the light there is always hope.'
(Unknown author)

The *Doppler Effect*, as was defined by Christian Doppler in 1842, originally describes the change in frequency and wavelength of a wave, when its source is moving relatively to a stationary *observer*. It is commonly experienced when a vehicle sounding a siren approaches, passes, and recedes from a stationary *observer*. Compared to the originally created frequency and wavelength, the received frequency, with corresponding changes in the wavelength, is higher during the approach, identical to the original frequency at the instant of passing, and lower during the recession.

The *Doppler Effect*, generally applies to any propagating wave motion originated from a given central source, and progressing towards or away from the *observer*.[1] It applies to the sound waves and also to the *progressing light*.

We have previously established that the speed of light propagating in vacuum is restricted by this medium to its constant value c. This has been already proven by *Willem de Sitter*, who used observations of a double-star system. He concluded that the speed at which the light propagates is independent of the movements of its source [2].

So far, there is nothing inexplicable about the light phenomenon. The light propagates similarly to the sound, and has its maximum speed limited by the restrictive medium. This limit on the maximum speed of propagating wave in a given medium affects only the wave, and it has no effect on the movement of its source. Theoretically, there is no limit to the speed of the source of wave, unless it is imposed by the medium. For better understanding we could compare the propagating source of light to the propagating sound source. Until the invention of supersonic aircraft it was not known what happens, when a flying plane exceeds the speed of sound. The first time the pilots attempted to exceed this speed they did not know what to expect. To their relief nothing sinister happened.

---

[1] More on this subject in appendix *A2*
[2] More on this subject in appendix *A3*

The pilot, overtaking the densely compacted sound waves, experiences a sonic boom, followed by a silence. The modulated sound of the engines of a plane, traveling faster than maximum speed of sound in the air, is left behind and the pilot cannot hear them.

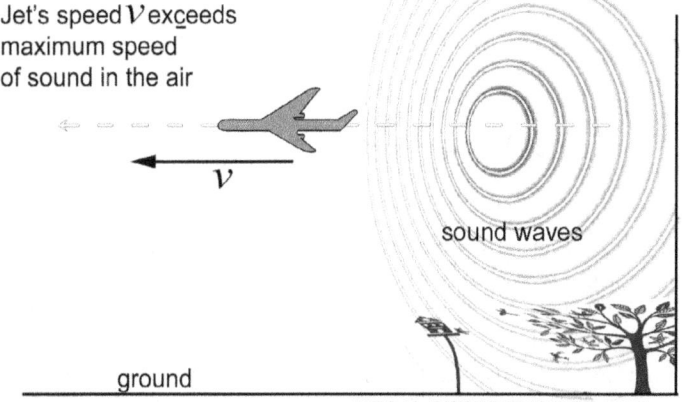

FIGURE 13.1 Jet flying with a speed exceeding the speed of sound in the air creates a sonic boom. After the boom, the sound waves will be forming behind the plane.
If the jet flies low, the sound waves will affect objects on the ground.

The aircraft or rocket, which could exceed the maximum speed of sound, is proof that there is no limit to the speed of the source of waves. That is, unless it is imposed by the medium. If the medium is the water, for example, it is obvious that any physical source of the sound would not be able to travel faster than the speed of sound in water.

There is also no limit to the relative speed of the source and *observer*. For example, should the *observer* move directly toward the light source, the *observed light* experienced by the *observer* will have a higher relative speed:

speed of light + observer's speed

The speed of *propagating light* and speed of *observed light* are crucial for the existence of life on our planet. The higher living forms have sensors, commonly called eyes, which are designed to cater especially for this small part of the electromagnetic spectrum.

The light reaching our eyes firstly falls on the cornea and lens of the eye, and only then proceeds to the retina, where it forms images of observed objects.

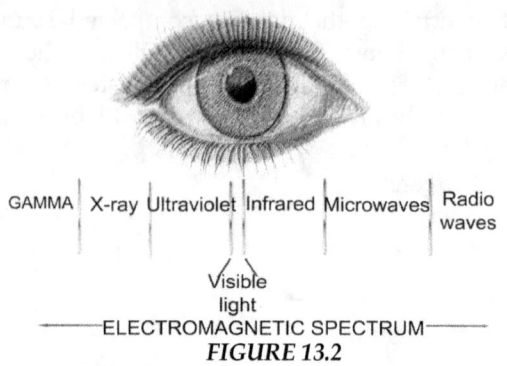

FIGURE 13.2

Only the visible part of the spectrum will reach the retina.

From these images our brain creates in our mind objects, which represent the world around us.

The eye filters the light in such a way that only its visible part is allowed to pass all the way to the retina. Light with invisible frequencies, like ultraviolet or infrared light, is absorbed by the cornea and lens, and never reaches the retina. Such filtering is destructive though, and it is interesting that both the cornea and lens could regenerate, but the retina cannot.

For us to see anything, the *propagating light* has to reach us with its constant speed $\underline{c}$. Should that not be so, then the resulting frequencies will be filtered off by the eyes.

Obviously, this is a protective design feature, and these properties could not have evolved. Without this protection, eyes of any living form will be destroyed, before they could modify themselves, as expected by evolution.

The space is still generally considered to be void of any filling substance. As we have already decided, we would ignore this argument entirely and accept that the vacuum restricts the speed of created light to the known constant speed $\underline{c}$.

Any true movement, i.e. not observed movement of the light, is relative to the vacuum. That implies that the vacuum, which actually represents the space, is rigid and forms what we have already defined as the *universal reference frame*.

We can conclude:

- *The human eyes were designed to cater for speed equal to or close to the constant speed of light $\underline{c}$ only.*

- *Any true movements of the progressing light is considered in relation to the rigid vacuum, i.e. to the universal reference frame.*

14. Some Uncorrected Errors
*'Any intelligent fool can make things
bigger, more complex, and more violent ... '*
(Albert Einstein)

Albert Einstein in his famous book, under headings '*The Lorentz Transformation*'[1], uses equations derived by Dutch physicist H. A. Lorentz.

Lorentz conducted an abstract experiment, where he sent a beam of light to a distant, stationary mirror. Firstly, from a stationary object and then from an object moving relatively to the mirror along a straight, parallel line. [2]

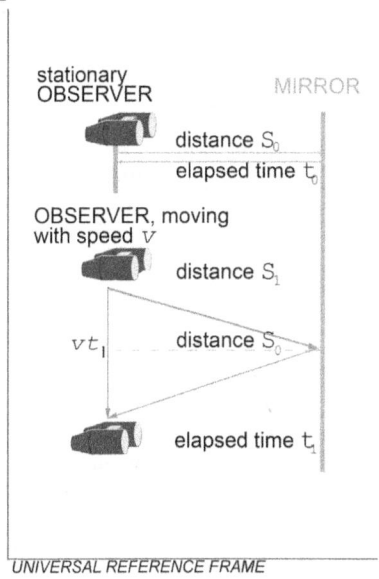

FIGURE 14.1 Lorentz's hypothetical experiment.

He then calculated the difference between the time taken for the light to return to the stationary, and to the moving *observer*. He concluded that it took longer for the light to reach the moving *observer* than the stationary *observer*. This is understandable, since the light reaching moving *observer* has to cover a longer distance.

---

[1] 'RELATIVITY THE SPECIAL AND GENERAL THEORY (translated by *Robert W. Lawson*, M.Sc. University of Sheffield), published in New York by *Henry Holt* and Company in 1920

[2] More on this subject in appendix *A4*

The result of these calculations is defined as the *Lorentz factor*. This formula includes time, constant speed of light c and the speed of moving *observer*.

It is incorrectly believed that Lorentz in his experiment calculated the change in the rate of time flow. Unfortunately, the whole section of nowadays science is incorrectly based on this assumption.

Truth is that Lorentz in his experiment did not calculate any change in the rate of time flow, simply because there is none. While the *observer* moves, its time progresses at its constant rate, not slower or faster.

Lorentz in his calculations actually assumed that the time flows at its constant rate, otherwise his experiment would not make any sense.

There is another incorrect assumption Lorentz did not take into account. It is the scenario, where the time delay is calculated for an *observer* traveling from a different position and/or in a different direction. For example, for an *observer* traveling toward the mirror, with increased *observer*'s speed the time the light travels is becoming shorter, not longer.

Compared to a stationary *observer*, in this scenario there is no time delay.

One interpretation of the special theory of relativity created the belief: *'on fast moving spaceships, time flows slower than on the Earth'*.

This belief is widely accepted, and even brought us the *'special theory of relativity time correction'*, which calculates how much slower a clock would run on a fast moving satellite, compared to the same clock placed on the Earth.

We have already proved that the light could be treated as a propagating wave in a medium, restricting the wave's maximum speed. In this concept the sound propagating in the air is similar and yet, it does not influence the rate of time flow. This erroneous assumption about traveling light could be further demonstrated in the example in figure 14.2.

In the universal reference frame, the *universal time* could be measured in many different ways: Sundials, water clocks, mechanical clocks, quartz clocks, atomic clocks, etc. To measure the *universal time* flow in this example, we selected a watch sending a light signal every second.

Both ibis and magpie could move in any direction and at any speed, even exceeding the maximum speed of light.

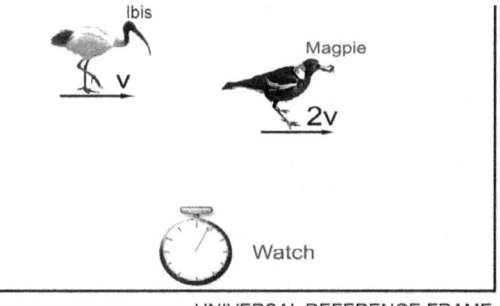

FIGURE 14.2 *In the universal reference frame, ibis moves with speed $\underline{v}$ and magpie with double speed $2\underline{v}$.*

If they move away from the watch with speed of the light or higher, the light signal announcing another elapsed second would not reach them. They just simply would not see the watch.

No matter how fast ibis or magpie moves, the flow of the *universal time* will not change for them, because the watch will always turn at the same rate. Independent, stationary *observer* in the universal reference frame will look at the watch, and will see that its rate has not changed.

The different situation arises, should the ibis and magpie set their time by observing the watch. Then the light signal, emitted by the watch, and received by them, becomes the *observed light*. There will be corresponding delay or speed up of the time the light signal reaches ibis and magpie. That will alter the time at which the light signals are arriving, and the experienced time will be different. The experienced time will be lagging behind the *universal time*, but the rate of flow of ibis and magpie's time will not change.

This experienced time will be their *subjective time* only. Obviously, ibis and magpie could have different *subjective time*, depending not just on their speed, but also on many other subjective factors, like their mental phase.

Should ibis and magpie have a pocket watch, set at the same rate of flow of the *universal time*, the watch will not slow down during their movement. The only way to make the watch go slower would be to mechanically change its settings.

This is a logical, and even practical proof that time does not slow down on a fast moving object. The assumption that a clock traveling at high speed will slow down is a pure fantasy.

Another overwhelming argument against the belief that fast movement causes slowing of the rate of time flow is actual misinterpretation of *Lorentz's* calculations.

Lorentz calculated only the delay in time needed for the light to reach a moving *observer*, instead of a stationary *observer*. What's more, the *Lorentz factor* applies only to the special situation, where the *observer* is moving on the straight line, perpendicular to the line connecting it with the light source and from a position closest to the light source. The value of the *Lorentz factor* for an object, moving in a different direction and from a different position, is different. [1]

Despite all that, the *Lorentz factor* is widely used in many calculations. One of them is a definition of the relativistic mass[2].

'Mass in special relativity incorporates the general understanding from the concept of mass–energy equivalence. The word, 'mass,' is given two meanings in special relativity: one ('rest mass' or 'invariant mass') is an invariant quantity, which is the same for all observers in all reference frames; the other ('relativistic mass') is dependent on the velocity of the observer. Roche states that about 60% of modern authors just use rest mass and avoid relativistic mass.' [3]

The *relativistic mass* was defined as:

Relativistic mass = rest mass x *Lorentz factor.*

$$m_r = m\gamma$$

One possible interpretation of this definition of relativistic mass has created the statement that *'nothing could travel faster than the light'*.

The *Lorentz factor*, as defined, increases with increasing speed of the moving object, and that will also apparently increase the object's mass to extremes. It is then believed that such enormous mass could not be moved.

For that I found the following explanation: [4]

'As an object approaches the speed of light, its mass rises ever more quickly, so it takes more and more energy to speed it up further. It can in fact never reach the speed of light, because by then its mass would become infinite, and by the equivalence of mass and energy, it would have taken an infinite amount of energy to get there.'

---

[1] More on this subject in appendix A4
[2] More on this subject in appendix A7
[3] *John Roche*, 'What is mass,' European Journal of Physics 2005 pp 239
[4] The international No.1 bestseller, 'A Brief History of Time,' written by *Stephen W. Hawking*, with introduction by *Carl Sagan*, reprint in August 1990

'For this reason, any normal object is forever confined by relativity to move at speeds slower than the speed of light. Only the light, or other waves that have no intrinsic mass, can move at the speed of light.'

This statement assumes there is a correlation between the speed of light, mass and energy of an object. This leads to more errors, because the relativistic mass is incorrectly used to calculate the famous Einstein's equation:

$$E = mc^2$$

To derive his equation, Albert Einstein used relativistic mass in purely mathematical calculations, interlocking energy, mass, and the maximum speed of light c.

Although I analyzed the Lorentz simple abstract experiment in many details, it does not reveal anything significant. *Lorentz's* calculations represent only what would be an experienced time delay, and not any change in the rate of time flow. What's more, the calculations are not complete and the derived formula is not precise.

It is impossible to find any logic in the definition of relativistic mass and there are not any reason for inclusion of this mathematical construct in modern physics.

To highlight the crucial points of this chapter, we could compare two typical definitions of *Lorentz factor*:

Lorentz factor as defined and used
$$\gamma = \frac{1}{\sqrt{1 - \frac{v^2}{c^2}}}$$

Lorentz factor derived for the *observer*'s movement along the direct, collision line with the light source:
$$\gamma' = \frac{1}{1 + \frac{v}{c}}$$

Both formulas are correct and differ only in conditions under which the measurements are taken.

Consider the following example in figure 14.3, illustrating two versions of *Lorentz factor*, each entirely appropriate for different *observer*'s movements. It is clear that any calculations including the *Lorentz factor* have to take into account all its versions, since even a slight change in *observer*'s position or direction of travel will produce completely different results.

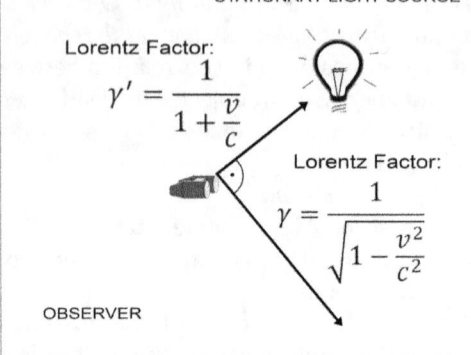

FIGURE 14.3 Observer moves toward the light source or by-passes it. For each situation, a different version of the Lorentz factor exists.

We can conclude:

- Any change in the rate of time flow cannot be calculated using the Lorentz factor. It is a mistake to believe that the time, reigning our world, i.e. the universal time, changes its rate of flow.

- Einstein has mistaken a simple time delay for slowing of the rate of time flow and as a result of that, his special theory of relativity, together with the concept of relativistic mass, is not correct.

- In view of the fact that relativistic mass is used to calculate the famous equation $E = mc^2$, it also applies that this equation is not correct.

> *It is humanly impossible not to make any mistakes. However, those making them should be judged not by the mistakes they made, but by their intentions and by their mistakes they tried to correct.*

15. Something to Think About
Everything should be made as simple as possible, but not simpler.
(Albert Einstein)

In the late 1920s, American astronomer Edwin Hubble derived from his calculations a conclusion that the universe is not rigid, but is expanding. He based his findings on *observed light*, coming from stars in distant galaxies. Most of the stars emit bright light, but some of them emit the red light, i.e. light, possibly shifted to the red part of the light's spectrum.

Hubble concluded that the shift toward the red color is caused by the *Doppler Effect*, and it signifies the star is moving away from us. On this conclusion he then built his concept of expanding the universe.

Although his findings and conclusions were generally accepted, considering the flaws found in the understanding of light and time so far, this subject is definitely not closed and should be investigated further. To start with, we would assume that Hubble is correct and the universe is expanding.

Taking into consideration the time taken for the light from a distant galaxy to arrive at our planet reveals already one major flaw of this theory.

For example, the Andromeda Galaxy is the closest large galaxy to the Milky Way, and can be seen unaided from the Earth. It takes 2.5 million years for the light from that galaxy to reach our planet. That means that any statements, based on observation of the light coming from that galaxy, refers to the situation 2.5 million years ago. The universe was expanding then, maybe, but is it still expanding now? Evidently, the red light emitted by some planet is not proof that at the present time the universe is still expanding. (Fact 1.)

To continue with our deliberations, we will simplify the whole situation and create a visual model, illustrated in figure 15.1-2.

The universe, filled with vacuum, is represented by the RUBBER, and the galaxies by the stationary ant and bee. They are surrounded by the vacuum, which expands with the expanding universe, i.e. with the expanding rubber.

Stretching the rubber sheet will cause the ant and bee to move relatively to the table and relatively to the center of the expanding universe. Relatively to the RUBBER, both keep their original positions, i.e. they stay on the same spot. Yet, the distance between them will increase from S_0 to $\underline{S}_1$.

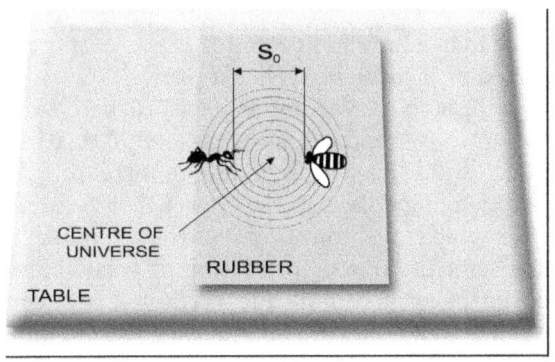

TABLE REFERENCE FRAME

FIGURE 15.1 *RUBBER represents our universe, consisting of vacuum. The ant and bee are representing galaxies, and relatively to the vacuum, both are stationary.*

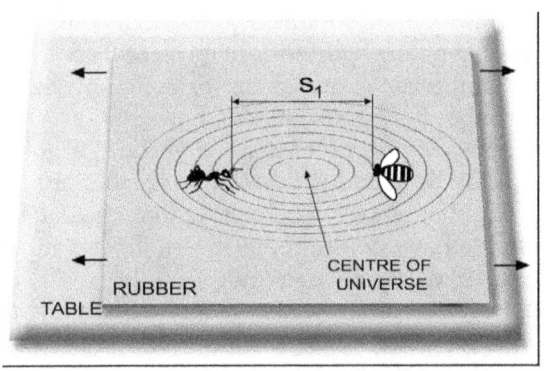

TABLE REFERENCE FRAME

FIGURE 15.2 *After stretching the RUBBER, ant and bee's positions relative to the table reference frame will change, but their initial position on the RUBBER remains the same.*

When the *bee* in figure **15.2** sends a light beam toward the *ant*, the situation with the *propagating light* is not entirely similar to the *Doppler Effect*. In *Doppler Effect*, the *ant* and the *bee* change their position relatively to the restrictive medium, which in our example is the *RUBBER*.

When stretched, both the *bee* and *ant* remain relatively to the *RUBBER* on the same spot, but relatively to each other, to the *CENTER OF UNIVERSE* and to the *TABLE* they are not stationary.

There are two different opinions, regarding the nature of the vacuum. More than a hundred years ago it was proposed that the vacuum is filled by a substance called *aether*.

The speed of light in the vacuum is 300,000,000 m/sec, in water only 225,000,000 m/sec and in glass even less, 200,000,000 m/sec. This limitation of the speed of light, due to increasing density of the speed-restrictive media, supports belief that vacuum is filled with the *aether*, which must consist of some material-like substance or containing some unknown, speed-limiting friction force.

The opposite belief considers the vacuum being completely void of anything. In this case there is nothing to limit the speed of light and the constant of the maximum speed of light would have to be one of the defined attributes of the light. It would have to be defined as a distance traveled by the light during a set period of time.

That would require specific definitions for the speed of light for different mediums, and also for their respective different densities. Obviously, the complexity of such a scenario, together with the fact that natural constants in our world are derived from the physical properties of our universe, rules out the possibility of maximum speed of light being defined.

We have now a logical proof that the vacuum is not void, and therefore it could be filled by the *aether* or it could be subjected to some form of radiation or force, limiting the speed of *propagating light*. Again, we have two different possibilities and obviously, from these two choices only one could be right.

Firstly consider the first choice, i.e. the vacuum filled with the *aether*.

The figure 15.2 supports this possibility and the *aether* could be actually represented by the RUBBER. It stretches with the expanding universe and takes with it both, the ant and the bee. Its presence, similarly to glass and water, changes the speed of light, which nicely explains the constant maximum speed $\underline{c}$ of light in vacuum.

It is then obvious that with the expanding universe, the *aether* would have to simultaneously expand as well, and when it does, it will lower its density, i.e. less mass would have to fill more of the space. That would logically change the ability of the *aether* to restrict the speed of light and therefore the constant speed $\underline{c}$ would have a different value than 3×10^8 m/sec.

To keep its speed limiting power, the *aether* would have to keep its constant density. To achieve that, there will be a need for more of the *aether* to fill the newly created space. In such a case, the *aether* would have to be created out of nothing, which makes the addition of 'more *aether*' meaningless. This logical flaw becomes obvious, when we substitute vacuum by water. Expanding a reservoir full of water will need more water to keep it at the same level. It is obvious that the possible existence of the *aether* presents a valid argument against the expansion of the universe.

(Fact 2.)

Inevitably, as the last option we should consider the vacuum being void, but exerting some force, which is slowing the speed of *propagating light*. Since the speed of light is slower in denser materials, this force could be similar to a friction force, with its strength increasing with the density of the medium.

Again, same as with the *aether*, it is obvious that with the expanding universe, the density of the medium will decrease, the limiting force will also decrease and the constant speed of light will then increase. This again represents a valid argument against the expansion of the universe.

(Fact 3.)

For the expanding universe scenario to be true, we have now excluded vacuum being completely void, we have also excluded being filled by material-like *aether* and the possibility of vacuum being exposed to some force is also excluded. Since we also have a proof that the speed of light is not defined, but it is limited by the vacuum, we can clearly state that the universe is not expanding. Furthermore, the vacuum is not void and is either filled by the *aether* or it exerts some frictional force on *progressing light*.

(Fact 4.)

Despite all that, we could still continue and consider some other, even more compelling reasons, for not believing that the universe is expanding.

One of such arguments is the fact that the expanding universe should also affect the Sun and all planets in our solar system. However, contrary to the principle of the expanding universe, the distances between planets in our galaxy are not changing.

The supporters of the expanding universe attribute this to a string of gravitational forces, existing between planets. It is believed that the expanding universe affects only galaxies, provided there are no gravitational forces between them.

Contrary to that, the theory behind the black holes in our universe claims that some distant stars create a strong gravitational field, and its effects are felt even on our planet. This opposes the belief of selective expansion.

The most compelling arguments for or against the expansion of the universe are the observed properties of the light emitted by the stars, given that they are the actual base for Hubble's calculations.

The light sent by the bee, and received by the ant, is not a simple *propagating light*, but it is the *observed light*. It will reach the ant with decreased frequency and speed, and quite possibly it won't be light at all, but some other form of electromagnetic radiation.
(Fact 5.)

We could persevere and consider yet another simplified scenario, illustrated in figure 15.3, where we have two objects, moving along a straight line and away from each other.

Planet $\underline{A}$ is moving away from the center of the universe, and so does the Earth. They both move with the same speed $\underline{v}$, representing the actual speed of the expanding universe.

The straight line, connecting planet $\underline{A}$ and the Earth, also passes through the center of the universe. Planet $\underline{B}$ is also moving away from the center of the universe with the speed $\underline{v}'$, but in a different direction. Based on the *Doppler Effect*, Hubble assumed that emitting the red part of the light's spectrum signifies the star is moving away from us. In such a case, the calculations derived from the *Doppler Effect*, if applied correctly, are valid and it should be possible to calculate $\underline{v}$, representing the speed of the expanding universe, measured relatively to the center of the universe

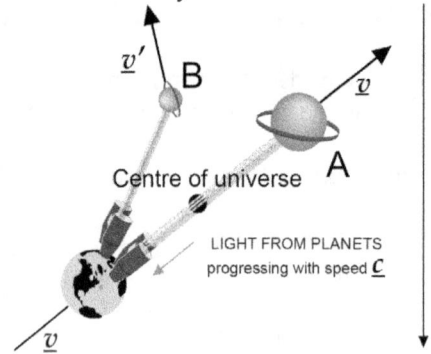

FIGURE 15.3 The expanding universe from its center with speed $\underline{v}$, relative to the universal reference frame.

Unfortunately, calculations used in the definition of *Doppler Effect* are incomplete. They are valid only for the source and *observer*, moving on a straight line, connecting them both with the center of the universe.

This is obviously not the situation with planet $\underline{B}$. Although it is moving away from the center of the universe, it is not moving along the straight line, connecting also the center of the universe and the Earth.

Therefore, any observation of this planet cannot be used to justify the expansion of the universe. Due to its speed and direction, planet $\underline{B}$ could emit a red light, and yet, the calculations derived from the *Doppler Effect* will produce different results. This possible scenario will produce different values for speed $\underline{v}'$, depending on the direction of the movements of the star and *observer*.

To evaluate the true speed of the expanding universe, only the calculations for planet $\underline{A}$ could be used. That means that only one selected planet or galaxy or some others very close to it, connected on a straight line with the center of the universe and the Earth could be used to calculate the approximate speed $\underline{v}$ of the expanding universe. Since we do not know where exactly the center of the universe is, we cannot be certain that the obtained measurements are correct.

As the last, and simplest argument presented here is the fact that the red light emitted by the planet could also simply indicate that in relation to the Earth, the planet is stationary and is emitting the red light. [1]   (Fact 6.)

Without knowing if the planet in relation to the Earth is moving, and if it does, in what precise direction and if through the center of the universe, we cannot calculate the speed of the expanding universe. Conclusion of this logical argument is that Hubble's calculations have a nihilistic chance of being correct.   (Fact 7.)

Based on these seven arguments we can conclude that at the present, there is no proof supporting the hypothesis of an expanding universe.

But there are still more arguments against the expansion and we should include the well known phenomenon, which is pertinent to this subject, namely the existence of black holes.

In the void universe cannot exist anything else being void, and still be different. Therefore, for the black holes to exist, either the universe or the black holes have to be filled by some tangible substance, be it *aether* or whatever.

[1] See appendix 6 for supporting calculations.

It is believed that such a black hole comprises a large star, whose gravity is so enormous that the light the star is emitting, cannot escape it.

To understand this phenomenon, we will try to create a black hole in our model. The following example illustrates a possible scenario:

FIGURE 15.4 Two-dimensional computer display has a hole, punched through it. This hole in our model could represent two-dimensional version of a black hole.

In figure 15.4, our computer display is just a thin sheet of light emitting elements, displaying an image, which exists in the two-dimensional universe of our model.

We could drill a hole through such a sheet, and suddenly we have a model with the two-dimensional equivalent of a black hole. *Tom* could walk across the screen and when he encounters the black hole, he would not see anything there. Any light beam sent to this black hole will never be seen again. The black hole represents an opening into a three-dimensional world, invisible and incomprehensible to *Tom*.

Should any object's coordinates 'fall' into this black hole, the object's image will be lost. Nonetheless, the object's code stored in the computer's memory will not be affected and obviously, drilling a hole through the display will not change the model's code.

This comparison between two- and three-dimensional black holes tempts us to draw conclusions that black holes in our universe are connecting us to the infinite world, where our mind and code exist.

We cannot even imagine what that all embodies, precisely as *Tom* and *Mary* in our model cannot imagine what that hole in front of them represents.

As in figure 15.4, black holes in relation to the expanding universe should actually expand with it, i.e. they should grow in size. Yet, it is generally believed that their expansion is not due to the expansion of the universe, but to the increase in their mass, attributed to the objects falling in it.

Recalling already quoted Galileo Galilei's words, *'two truths cannot contradict one another'*, forces us to decide between expanding and non-expanding universe.

Against believing in the expanding universe are all, so far derived facts. But there is another, more compelling argument:

the universe is not infinite, as would its infinite expansion require. As we already proved in chapter 3, the speed of light is not defined in the *infinity*: $\underline{D}_{Light} = t \cdot c_{Light}$ *(time traveled x speed of light)* [1]

In *infinity* $t = \infty$ therefore $\underline{D}_{Light} = \infty$
This equation becomes $\infty = \infty \cdot c_{Light}$
or $c_{Light} = \infty / \infty$ which is undefined.

Should the universe be infinite, then the speed of light is not defined and the light itself cannot exist there.

However, light exists in our universe and therefore the universe cannot be infinite. Obviously, this fact alone does not allow the infinite expansion of the universe.

Now we have another limitation to the expansion of the universe: If the universe expands, then it cannot expand indefinitely.

We can conclude:
- *Derived logical arguments do not support expanding the universe. The stars, planets, and whole galaxies are either stationary or moving relatively to the universal reference frame.*

- *The presence of the light in the universe signifies that the universe is not infinite. This would limit its possible expansion and should the universe expands, it cannot expand indefinitely.*

- *Black holes are most likely the 'windows' into a different world of our mind, called infinity.*

[1] See paragraph 3.

Book's Summary
'Cogito, ergo sum.' - I think, therefore I am. - (Renè Descartes)

Considering our world as an image in the mind of an *observer* existing in *infinity* explains nicely the paradox of something existing only if it is observed. It also explains why something could exist simultaneously in more than one place (quantum superposition), and why any changes in the object's image reflect immediately on all instances of that object (quantum entanglement). This modeling also allows for the existence of parallel universes.

The objects in our world are three-dimensional images, formed in the mind of an *observer* existing in the infinite world. That makes our world three-dimensional and therefore it cannot be infinite, as explained in paragraph 3.

Our brain is a part of our existence in this world and connects to our mind. Small, active part of our brain is used to control bodily functions, create feelings and individual intellect. It makes us what we are. The larger, passive part of our brain contains only addresses, where all the different information is stored in our mind.

Our mind represents our existence in the unknown world, where it forms a part of our *observer's* mind. Contrary to a limited capacity and limited life of our brain, our mind and the mind of our *observer* in *infinity* are both limitless.

Gathering information in our mind during our life is the main purpose of our existence. Mind's capacity is limitless, but retrieving stored information is restricted by the capacity of our brain, i.e. how many addresses it can contain.

During our lives these addresses could be removed or replaced, thus creating the process of remembering and forgetting.

Besides the *universal time*, there is also another time in our world called *subjective time*, experienced by each individual and they are both independent of each other. The rate of flow of *universal time* is constant throughout the whole universe and the rate of flow of *subjective time* is subject to many factors affecting the individual mind, such as the *observer's* surroundings, frequency of experiences, mental health, etc.

Objects exist simultaneously in our world and in the *infinity* and they have infinite minds.

Since the present component of the *universal time* is infinitely small, it could exist only in the *infinity*.

Therefore, objects existing there are not just limitless, because they are infinite, but also they are eternal, because they exist in the never-ending present.

When our brain is awake, accessing information follows the direction from the active brain stack to our mind. When in induced subconsciousness, our brain communicates directly with our mind. In that stage we could recall long forgotten information, still stored in our mind. The subconsciousness could be induced by drugs, hypnotism, during sleep, etc.

Our body is a very cleverly designed machine, self-built, self-repairing, and self-controlling many of its functions. It is not self-programming, though.

It digests organic matter and converts it to hydrocarbons. They are delivered by blood to muscles and other organs, where they burn and turn into energy.

We have only a partial control over our body, which could at any time disregard our wishes and independently perform whatever is for the given circumstances programmed. Our body and our brain enable our mind in *infinity* to be present in our three-dimensional world.

We have no concrete proof of the presence of the hidden word we call *infinity*, but by observations and modeling we could ascertain its existence.

There exist laws of nature, which we cannot change and the same applies to hidden constants, for example the Planck's and gravitational constant. They are defined by the structure of the universe and we have no means of changing them. They are part of the underlying matrix of our world, designed not in our world, but in the infinite world.

The creation of our world follows a typical procedure of programming robots in our world, already known to us. It starts with some simple objects, followed by more complex and advanced objects. The steadily progressing miniaturization of the objects' design is an obvious sign of increasing sophistication of the advancing technology of our civilization.

When we compare this trend to our planet's development, the similarities become very obvious. Firstly, only very primitive life started to appear, being later replaced by better designed forms. Unsuitable life forms, for example prehistoric dinosaurs, were 'deleted', since they presented a hindrance to further development of more versatile forms, like for example, the humans.

Even the initial development of humans was not what we are now. Firstly, they existed as primitive types, which later were replaced by more sophisticated forms.

Since the same code was reused and refined, we find many similarities between species, like for example between fish and birds or Neanderthal man and Homo sapiens.

Obviously, everything in our world is programmed, all follows the same in-built pattern of 'boom and bust'. We are similar to bacteria cultures and short of total destruction of our civilization, we cannot find a common consensus of how to break this disastrous cycle.

For humans, light is the most important attribute of our world. Our eyes were actually designed to cater for exactly the frequency and wavelength of this visible part of the electromagnetic spectrum. That correlation has definitely not evolved, but was programmed.

In this book we established that light travels with its constant speed only in relation to the medium in which it propagates and is independent of the speed of its source. The general belief that 'nothing could travel faster than light,' was found misguided. It is based on the calculations involving the *Lorentz factor*, which is used to define the concept of relativistic mass. The *Lorentz factor* describes only the delay in time for the light to reach a moving object instead of a stationary object. It does not represent any change in the rate of time flow, as is wrongly assumed.

The belief in slowing down the rate of time flow during interplanetary voyages based on the *Lorentz factor* is incorrect and cannot be substantiated.

The *progressing light*, observed by different *observers*, becomes *observed light*, which is different. The *observed light* could have a shifted frequency and it could even have a different speed than its constant speed c.

The concept of *observed light* breaks down the principle of relativity, formulated by Galileo. This principle does not apply to the *observed light*, experienced by the moving *observer*, since there is a difference in frequency and speed between the *progressing light* and *observed light*. This is caused by movement of the *observer* in the restrictive medium.

The aim of teaching at our educational institutions is to present students with some already derived conclusions. During the course they are required to learn these accepted conclusions and later be examined.

They are not required to analyze them and most of the students will not be doing it, even after leaving the educational institutions. This process thus carries many errors, introduced into the science, even a long time ago.

Examples in this book depict the misunderstood Lorentz Transformation, and mistakes created by using it incorrectly; especially in Einstein's special theory of relativity. That has caused an inaccurate understanding of our world and universe. That has a direct impact on technological development, like for example is the use of the 'flexible time' concept.

Our programming model has simplified our perception of the universe. The logic of our model makes it easier for us to understand the beginning of our universe and also its possible ending.

It is obvious that in our three-dimensional world, we will be looking in vain for such an answer. We are exactly in the same situation as are the objects in our two-dimensional model, searching in vain for the answer to the same question in their two-dimensional world.

We, as the creators of our model, are the only ones who know the answer to their question. Then our question, concerning our existence in our world, could only be answered by the creator of our three-dimensional world.

Our world is a sophisticated computer simulation, and during our presence here, we accumulate experiences, which we take with us into our existence in *infinity*.

At the end of our lives, the reason why we are here becomes plainly obvious to most of us:

We are here to learn and become a better persons than we were at the start of our lives.

APPENDICES

Appendix 1: Numbers and Our Model

'Mathematics is the language with which God has written the universe.' (Galileo Galilei)

Mathematics is obviously the underlying matrix of our world. We already know it exists and to an inquisitive mind it reveals itself. It involves mathematical entities, like for example, basic calculations and numbers. In our civilization numbers have a long history and are with us for thousands of years..

The very first numbers were used for counting physical objects: 10 fingers, 100 coins, 5 trees, etc. They were called the natural numbers. Later the number 0 was included in this series of numbers and they were called whole numbers (0, 1, 2 ... 124, ...). After the negative numbers were introduced, all these numbers were called integers (... -3,-2, -1, 0, 1, 2 ...).

These numbers could be added, subtracted, and multiplied. When divided, they formed fractions, i.e. a ratio of two numbers. With fractions included, all these numbers are called rational numbers.

Before Greek mathematician Hippasus[1] came to the scene, it was believed that all these numbers fill the numeric line fully.

Number Line

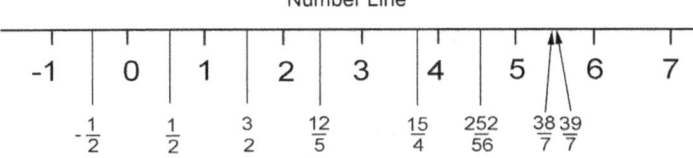

Some five centuries B.C., a group of followers of Pythagoras, including Hippasus, gathered aboard a ship to discuss the categorization of the number √2, discovered by Pythagoras.

√2 is called square root of two and it is a number, which multiplied by itself gives the number 2.

(√2 = number $\underline{a}$, where $\underline{a} \times \underline{a} = 2$)

Since √2 is represented by the length of hypotenuse of the right angle triangle with two sides equal to 1, it was believed that it exists on the number line and therefore could be expressed as a fraction of two numbers.

---

[1] This is a free interpretation of topic taken from the book, 'The Bedside Book of Geometry,' by *Mike Askew* and *Sheila Ebbutt*, publisher PIER 9, ISBN 978-1742660363)

That was the case until Hippasus produced his brilliant reasoning and proved that √2 could not be expressed by a fraction. His proof is quite simple:

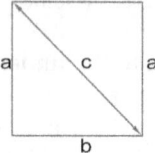

If a = b = 1, and they form a square, then according to Pythagoras, c = √2. If √2 could be expressed as a fraction, then where x and y are any two numbers, which have no common factors. The following must apply:
$$\sqrt{2} = \frac{x}{y} \quad 2 = \frac{x^2}{y^2}$$

x^2 is an even number, since it is a multiple of 2: $\quad 2y^2 = x^2$

Then x has to be also even, since squaring odd number will never produce an even number. Therefore, x could be expressed as $2z$, where number z could be any number. Multiplied by 2 will always produce an even number. Then the following applies:

$$2y^2 = x^2 \implies 2y^2 = (2z)^2 \implies 2y^2 = 4z^2 \implies y^2 = 2z^2$$

The number y is then also an even number, what makes both numbers x and y even numbers and they both have number 2 as a common factor. When these numbers were selected, it was assumed they do not have common factors. Given that this is not the case, √2 could not be expressed as a fraction of two numbers, which is the proof by contradiction.

All Pythagorean followers, taking part in the discussion on that boat believed that all numbers could be expressed as a fraction. They could not prove Hippasus being wrong and probably for that reason alone they threw him overboard and he drowned.

All the numbers on the number line, which could be expressed as a fraction, are categorized as *rational numbers*. The other numbers, i.e. those which could not be expressed as a fraction, are called *irrational numbers*. Both categories, i.e. rational and irrational numbers are called real numbers. Besides real numbers we also have other categories, but since they are not used in this book, we will not elaborate on them.

Objects in our model could be programmed in such a way that they would know how to add and subtract. This could be represented on the screen by a simple abacus, where a number of beads are added or subtracted.

Both of these calculations should be clearly seen by other object's displayed images, provided we add seeing capability to these objects.

Therefore, we could be satisfied that these two calculations are fully represented in our model. Unfortunately, the use of abacus is limited in its representation of numbers and should we require inclusion of some other, more involved calculations, we have to use some other tools.

We could start with an abstract number line, where numbers are represented by a distance from the point representing the number zero.

In our three-dimensional world, only three-dimensional objects exist and since the number line is only one-dimensional, the only place where it could exist is in our mind - we could only imagine such a line.

Part of such a line could be represented in our real world by a school ruler and in our model, which is only two-dimensional, we could use a picture of such a ruler. The ruler contains only a small subset of numbers, though. We could find there numbers 0, 1, 2, etc., and the ruler usually ends with number 30.

Most rulers also have fractions, for example 1.5 or 2.1. All these numbers could be expressed as a fraction of two whole numbers
(1.5 = 3/ 2, 2.1 = 21/10).

These are the already mentioned rational numbers and the irrational numbers, with their never ending number of decimal places, should be somewhere on the number line too. For example, the number is √2 should be somewhere between the numbers 1.41 and 1.42:

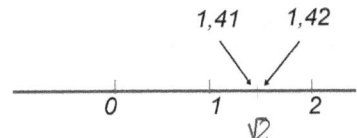

FIGURE A1.1 Number line and number √2.

Most electronic calculators would display this number as 1.4142, but if you have more advanced calculator, this number could be 1.414213562.

Although the number of decimal places of √2 never ends, this number should be situated on the number line and should be represented by the length of the line, somewhere between 0 and 2. This number is actually situated somewhere between 1.4142 and 1.4143, more precisely between 1.414213 and 1.414214, and even more precisely, between 1,414213562 and 1,414213563, and so on. In this manner we could continue infinitely, meaning that the number √2 must be in *infinity*.

Yet, we could be easily tricked to believe that irrational numbers exist in our three-dimensional world. For example, already mentioned √2 in figure 1.2 is represented by a definite length.

The triangle drawing is only two-dimensional and any individual line is only one-dimensional. In our three-dimensional world they exist only as images and we could never measure the precise length of such an image of a line. However, we could calculate, to some degree of accuracy, the length of the line <u>*AC*</u>.

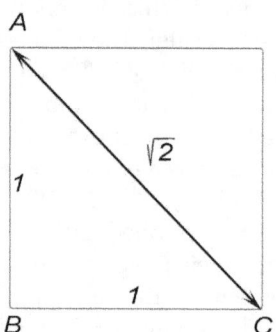

FIGURE. A1.2 *The distance AB = 1, BC = 1,*
angle ABC is 90°
and √2 represents numerical value of distance AC.

According to Pythagorean Theorem, the hypotenuse is equal to √2:

$$AC^2 = AB^2 + BC^2$$
since $AB = BC = 1$, then $AC^2 = 1^2 + 1^2 = 2$
therefore $AC = \sqrt{2}$

That length must exist somewhere, since it represents the length of the hypotenuse, i.e. the distance between two points. The only place where it could exist is our mind, and only in our mind we could multiply the distance √2 x √2 and obtain number 2.

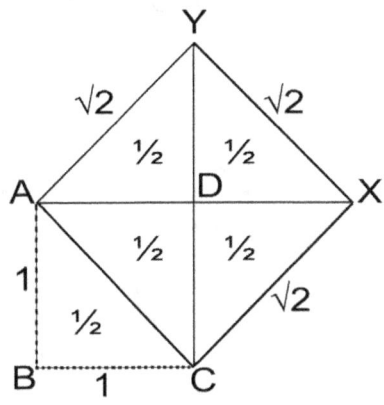

FIGURE A1.3 √2 *is represented by distance AC.*

The area of rectangle ABCD	AB x BC = 1
The area of triangle ADC	1/2
The area of rectangle ACXY	4x 1/2 = 2
Therefore	AC x CX = √2 x √2 = 2

What we just did, was to multiply two values existing in *infinity*, and therefore *infinity* must also exist in our mind, which must be then also infinite.

For our model to truly represent our world, it would have to also contain the *infinity*. Unfortunately, this is not possible since computers in our three-dimensional world have limited storage capacity.

All we could do is to mimic the *infinity* by some constant function or symbol, and that way create an apparent *infinity*, which should be adequate for the purpose of our model.

In our three-dimensional world, the real *infinity* we encounter on the number line cannot be substituted by any function, since we have no access to the code of the program, which created our world. Therefore, this *infinity* must exist in a different world than ours.

That implies that our mind, which contains *infinity*, must also exist in a different world, despite the fact that we exist in our three-dimensional world.

We can assert that our mind does not exist in our world, but in some other, infinite world, and therefore our mind is infinite.

Appendix 2: The Doppler Effect

Insanity is doing the same thing, over and over again, but expecting different results" (Albert Einstein)

Well before the introduction of the special theory of relativity, it was known that the movement of the source of waves through some medium affects their frequencies and wavelength. This phenomenon was named the *Doppler Effect*:

'The Doppler Effect or Doppler shift, is the change in frequency of a wave (or other periodic event) for an observer moving relative to its source. It is named after the Austrian physicist Christian Doppler, who proposed it in 1842 in Prague. It is commonly heard when a vehicle sounding a siren or horn approaches, passes, and recedes from an observer. Compared to the emitted frequency, the received frequency is higher during the approach, identical at the instant of passing by, and lower during the recession.

When the source of the waves is moving toward the observer, each successive wave crest is emitted from a position closer to the observer than the previous wave. Therefore, each wave takes slightly less time to reach the observer than the previous wave. Hence, the time between the arrivals of successive wave crests at the observer is reduced, causing an increase in the frequency. While they are traveling, the distance between successive wave fronts is reduced, so the waves 'bunch together.' Conversely, if the source of waves is moving away from the observer, each wave is emitted from a position farther from the observer than the previous wave, so the arrival time between successive waves is increased, reducing the frequency. The distance between successive wave fronts is then increased, so the waves 'spread out'.[1]

The statement from Wikipedia is valid only, if there is a medium in which the waves propagate and which restricts their maximum speed. These criteria apply, for example to sound waves propagating through the air.

The sound, generated by the train's whistle, must travel through the air, which is stationary and limits the speed of sound c_s to maximum speed in dry air at 200C to 343.2 m/sec. This causes the waves, sent by a moving source, to bunch together in front and the waves at the back to spread out.

[1]https://en.wikipedia.org/wiki/Doppler_effect (JULY 2015)

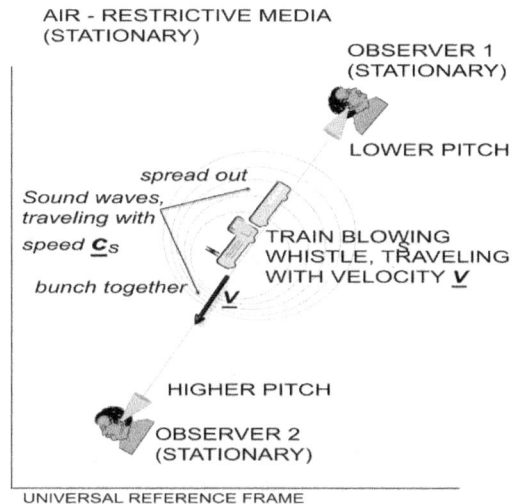

FIGURE A2.1 *The train moves and blows a whistle. It passes OBSERVER₁ and approaches OBSERVER₂. Both observers are listening to the sound emitted by the whistle. OBSERVER₂ hears a high pitched tone and OBSERVER₁ low pitched tone. OBSERVER₁, OBSERVER₂ and the air are stationary.*

Without a restrictive medium the combination of the speed of sound wave $\underline{c}_s$ and the speed of the train $\underline{v}$ should give us the resulting speed of propagating sound waves: $\underline{c}_s + \underline{v}$ in front, and $\underline{c}_s - \underline{v}$ at back.

With a restrictive medium involved, the speed of sound $\underline{c}_s$ is completely independent of the speed of the train.

It will reach a stationary *observer* with its constant speed $\underline{c}_s$, regardless of the speed of the source $\underline{v}$.

This scenario is illustrated in the figure A2.2, where the source is moving and the *observer* is stationary. The initial, unchanged wavelength of progressing sound is $\underline{\lambda}_0$ and frequency f_0.

The modified wavelength reaching the stationary *observer* is $\underline{\lambda}_1$ and frequency f_1. In time $\underline{t}_1$ the sound's source travels distance $\underline{vt}_1$.

With each emitted wave of frequency f0, the source will travel the distance $\underline{vt}_1$, where:
$$f_0 = \frac{1}{t_1} \Rightarrow t_1 = \frac{1}{f_0}$$

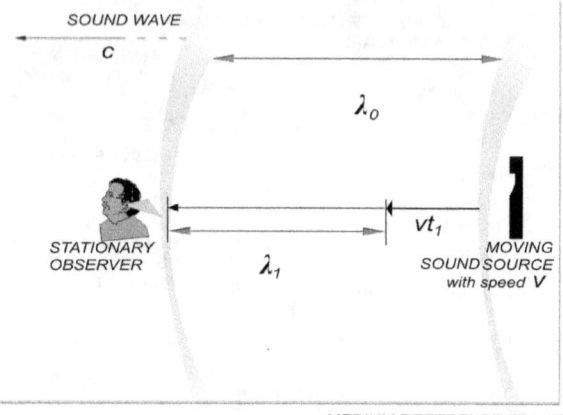

FIGURE A2.2 *The air restricts the maximum speed of sound to c_S. In MEDIUM REFERENCE FRAME, the sound source is moving on a straight line toward the stationary observer with the speed $\underline{v}$.*

The modified wavelength is: $\lambda_1 = \lambda_0 - vt_1 = \dfrac{c_s}{f_0} - v\dfrac{1}{f_0} = \dfrac{c_s - v}{f_0}$

The change in frequency, caused by the *Doppler Effect*, is due to the moving source of sound. The frequency of the sound emitted by moving source and reaching stationary *observer* is then f_1:

$$f_1 = \frac{c_s}{\lambda_1} = \frac{c_s}{\frac{c_s - v}{f_0}} = f_0 \left(\frac{c_s}{c_s - v}\right)$$

For the source moving away from the *observer*:

$$f_1 = f_0 \left(\frac{c_s}{c_s + v}\right)$$

For these calculations to be valid, the source has to move on a straight line, connecting it with the *observer*. Should the *observer* be positioned away from this line, these calculations cannot be used.

Although the *Doppler Effect* did not provide any calculations for such a situation, the *observer* would still experience a shift in frequency. As the train approaches, the frequency will be continuously increasing and when departing from the *observer*, the frequency will be continuously decreasing. Unfortunately, to calculate the resulting frequency for this situation will require more complex calculations.

(The calculations involving *observer's* movement, which were not part of the original *Doppler's effect* definition, are described in appendix 5.)

It is also important to realize that the depicted scenario for calculation of effective frequency deals with two-dimensional scenes, whereas our world is three-dimensional and therefore these calculations have limited use.

We can conclude:

- *The formula for Doppler Effect has limited usage and could only be applied in purely theoretical scenarios and should not be seriously considered for anything else.*

Appendix 3: The Michelson–Morley Experiment
"I always tried to turn every disaster into an opportunity." John D. Rockefeller

Modern science is very uneasy with any mysteries and does not wish to accommodate their existence. Yet, one mystery is still with us and what's more, it was created by science itself!

Towards the end of the 19th century, the scientific community was divided in two camps: One believing that the universe is filled with a substance called *aether* and the opposing camp believed the universe is filled with nothing. There was no proof supporting either belief and finally, in 1887, two English scientists *Michelson* and *Morley* decided to eliminate this uncertainty, once and for all.

Fig. A3 Michelson's interferometer.

They were encouraged by the latest discoveries involving the light, and decided to use it to confirm the presence or the absence of the *aether* in the universe.

They used an apparatus with two arms, forming a right angle between them. At their intersection was a prism, separating the beam of generated light in two beams and sending each beam to the mirrors fixed at the end of each arm. The expectations were that the returning beams will form a light pattern on the display, observable through the eyepiece.

During the experiment, the arms were slowly rotated, as illustrated on the following diagram:

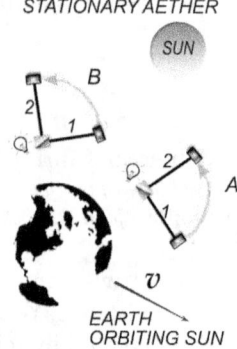

Fig. **D2** *The rotating interferometer stationed on Earth.*

The general consensus was that ideally, both light beams arrive at the eyepiece at the same time. In the presence of the *aether*, the longitudinal beam sent in the direction of the moving Earth will be exposed to an '*aether wind*', and it will encounter greater *aether* resistance, and should therefore arrive at the display slightly later than the transversal beam.

In the figure **D2** are illustrated two positions, **A** and **B** of the instrument. In the position **A**, the arm **1** is pointing longitudinally in approximate direction of the orbiting Earth, i.e., against the apparent *aether wind*. The created resistance will slow down the progressing light on arm **1** more than on arm **2**. It was expected that in the first position **A**, the beam on arm **1** should arrive at the display slightly later than the beam on arm **2**.

This discrepancy in times of arrival of both beams would then affect the light pattern, formed on the display and observed through the eyepiece. To ascertain that the arm **1** points in the same direction as the moving Earth, the arms of the instrument were rotated, while the displayed light patterns were observed.

At some point during the rotation, arm **1** would be definitely pointing in the same direction as the moving Earth, and arm **2** would be pointing directly to the Sun. At that point, the possible resistance of the *aether wind* would be greatest. That should be seen on the reflected light pattern as a distinct interference of two waves, arriving at slightly different times.

The instrument was turned even further to position **B**, and after every changed position the resulting observed interference should have been obvious. At least that was generally believed and expected. To a great surprise, no interference was observed and the conclusion was that both light waves arrived, more or less, at the same time:

'But when the experiment was made, it was found that the two beams arrived back at the same time. ... Now it must be recognized at once that this was a most extraordinary thing. Here was an experiment, performed with every care and apparently with full understanding of what was being done, which completely failed to give the result that common sense would have thought inevitable.

... If any explanation is to be given, therefore, it must necessarily involve something revolutionary.' [1]

Evidently, something 'extraordinary' and 'revolutionary' always deserves a sensible explanation. Since the whole situation is part of our physical world, all relevant physical laws must apply. Then the convincing justification, in some understandable and logical terms, for the obtained results has to be found.

At the time of the experiment, *Michelson*, *Morley* and all the others in the scientific community, based their conclusions on a simple, common sense example, where a body traveling against the wind will be slowed down by the wind more than the body traveling in a transverse direction.

That, for example, would be true for the sound traveling through the air, due to the physical nature of the sound waves and the propagating medium. For the same reasons probably, it was wrongly assumed that:

- The light progressing in longitudinal direction, i.e., against the *aether wind*, would take longer to return to the eyepiece than in the transversal direction.
- Without the *aether wind* , both beams arrive at the eyepiece at the same time.

[1] Described by *Herbert Dingle* in 1922, in his book 'Relativity for All'.

Based on the apparent failure of this test, the conclusion was that there is only a vacuum filling the universe and no *aether*. This conclusion is still considered as valid even now, and the presence of the *aether* in the universe is not generally accepted.

At the time of the experiment it was believed that the speed of light will be affected by the friction created by light moving through the *aether*, i.e., the speed of light will slow down.

It was also assumed that the Earth and the instrument are both moving with the same speed $\underline{v}$, relative to the universal reference frame, and the Earth's orbiting speed $\underline{v}$, which is approx. 30 Km/sec, is used in calculations.

Corrections to original calculations were later made by *Alfed Potier* and *Hendrik Lorentz*, which proved that both beams do not arrive at the eyepiece at the same time, as was originally presumed. The corrected calculations proved that even in the absence of the *aether*, for the constant speed of light $\underline{c}$, the longitudinal beam always arrives later than the transversal. The whole experiment and all the corrected relevant calculations could be presented in a simplified, corrected scenario as illustrated by the following figure.

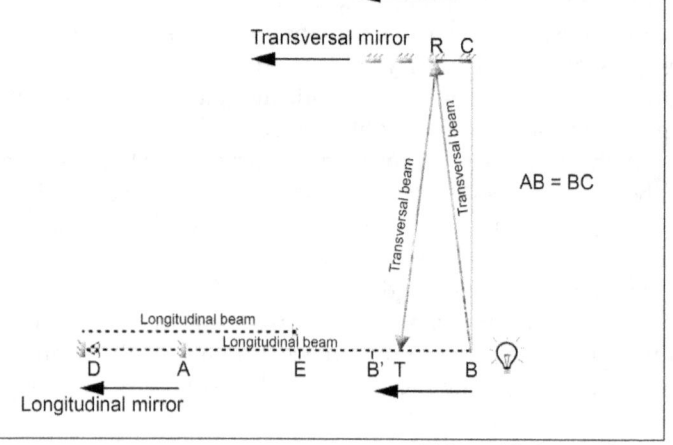

Fig. D3 *Corrections done to Michelson-Morley experiment. The longitudinal beam of light from the source will travel from $\underline{B}$ to $\underline{D}$, and then reflects to $\underline{E}$. Before reflecting, the starting position $\underline{B}$ moves left by $\underline{AD}$, to position $\underline{B'}$, and after reflecting moves to position $\underline{E}$. The transversal beam will also start at $\underline{B}$, travel to $\underline{R}$ and then reflects to $\underline{T}$. Both mirrors, the light source and the eyepiece will move with speed $\underline{v}$.*

In figure **D3** the traversal beam is sent to traversal mirror in time $\underline{t_1}$ at speed $\underline{c}$ covers distance $c\underline{t_1}$ = BR => $\underline{t_1}$ = BR/c

Since BC = AB

$$BR = RT = \sqrt{RC^2 + BC^2} = \sqrt{RC^2 + AB^2}$$

$$t_1 = \frac{\sqrt{RC^2 + AB^2}}{c} > \qquad t_1^2 = \frac{RC^2 + AB^2}{c^2}$$

Since RC = $v\underline{t_1}$

$$t_1^2 = \frac{(vt_1)^2 + AB^2}{c^2}$$

$$t_1^2 c^2 - v^2 t_1^2 = AB^2$$

$$t_1^2 = \frac{AB^2}{c^2 - v^2} > \qquad t_1 = \frac{AB}{\sqrt{c^2 - v^2}}$$

Total transversal time $\underline{t_T}$ = 2 $\underline{t_1}$ => $t_T = \dfrac{2AB}{\sqrt{c^2 - v^2}}$

Time for the longitudinal beam to travel consists of two time intervals $\underline{t_1}$ and $\underline{t_2}$.

$$t_1 = \frac{BD}{c} = \frac{AB + vt_1}{c} \Rightarrow ct_1 = AB + vt_1$$
$$t_1(c - v) = AB \Rightarrow t_1 = \frac{AB}{c - v}$$

$$t_2 = \frac{DE}{c} = \frac{AB - EB'}{c} = \frac{AB - vt_2}{c} \Rightarrow ct_2 = AB - vt_2$$

$$t_2 = \frac{AB}{c + v}$$

Total longitudinal time $\underline{t_L}$ = $\underline{t_1}$ + $\underline{t_2}$

$$t_L = \frac{AB}{c - v} + \frac{AB}{c + v} = \frac{AB(c + v) + AB(c - v)}{(c + v)(c - v)}$$

$$t_L = \frac{ABc + ABv + ABc - ABv}{c^2 - cv + vc - v^2} = \frac{2ABc}{c^2 - v^2}$$

In the graph in figure **D4** the distance **_AB_** was set to 9 m[1], since that was the original length the light traveled. The full line represents the longitudinal movement, with corresponding values of **_t_1_**, the time of arrival of the longitudinal light beam. The dotted line represents the transverse movement, with corresponding values of **_t_2_**, i.e., the time of arrival of the transverse light beam.

Both precise moments of arrival of each light beam depend on the speed **_v_**, at which the source of the light, stationed on the Earth, is moving through space. For low speed **_v_** there will be only a negligible difference between **_t_1_** and **_t_2_**, and only at higher speeds the difference becomes noticeable.

Since the speed of orbiting Earth's is 30,000 m/sec, the time and distance traveled by longitudinal beam:

time taken	6.0000035266687399e-8 sec
distance traveled	18.000010580006219 m

Transverse beam:

time taken	6.0000017633341110e-8 sec
distance traveled	18.000005290002331 m
Difference time	1.7633346289845457e-14 sec
distance	5.2900038873815447e-6 m

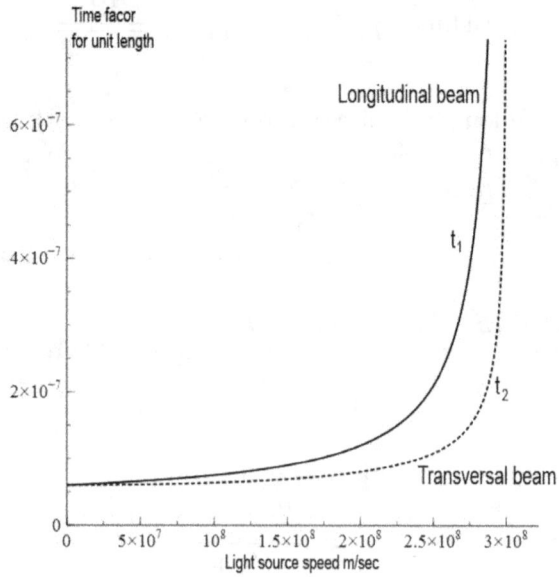

Fig. D4 Michelson-Morley experiment.

[1] For more precise results the beam was reflected across the arms more than once. For a greater simplification, we used 9 m length only.

It is obvious that even without any interference from the *aether wind,* it takes longer for the beam of light to return from longitudinal direction than from transversal direction, i.e., the beam from transverse direction arrives sooner than the beam from longitudinal direction.

This result, produced one year later by corrections done to the initial calculations by *Alfed Potier* and *Hendrik Lorentz,* supports the original idea that two waves arriving at different times should produce interference patterns on the display. Yet, the results of observations did not support these expectations, since during the experiment there was no interference observed.

Michelson expected that Earth's motion would produce a pronounced shift in light pattern, but in the published results of this experiment, the greatest separation they achieved was only 0.018 fringes.

In the experiment, the speed of orbiting Earth $\underline{v}$ was considered to be 30 000 m/sec., but in reality this speed does not represent the real speed of the Earth in space. It is believed that the speed of the solar system in the universe is 200 Km/sec, which was not included in the original calculations, since such an increase would have an insignificant impact on the calculated results.

For the apparatus arms' length of 9 m, the difference between distances travelled by transversal and longitudinal beams represents approx. 10 of a typical light wavelength of light, i.e.,500 x10^{-9} m. For greater accuracy, *Michelson-Morley* later reflected each beam many times before combining them on the eyepiece. That would make the difference in the distance travelled by each beam grater and it would easily accommodate some additional beams sent by the light source.

Shall we change the length of both arms to, for example, 90 m, we will obtain a slightly different results.

Longitudinal beam:
               time taken   6.0000060000060005e-7 sec
               distance traveled 180.00018000018002 m
Transverse beam:
               time taken   6.0000030000022500e-7 sec
               distance traveled   180.00009000006750 m

Difference   time   3.0000037505329433e-13 sec
                    distance   9.0000112521693154e-5 m

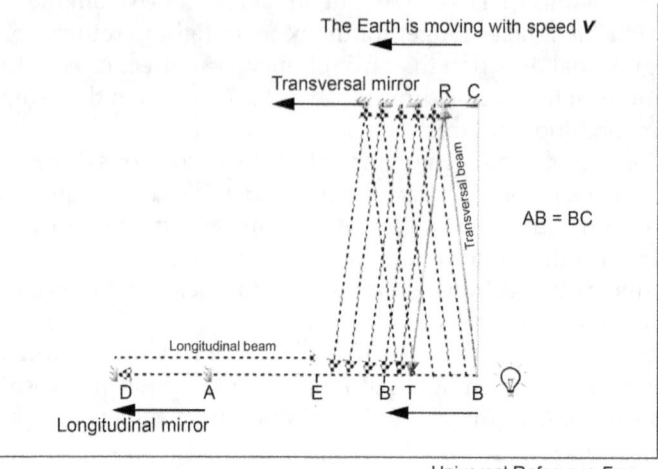

Fig. D5 Michelson-Morley experiment.
Before the longitudinal beam arrives at the eyepiece, additional transversal beams could arrive at the eyepiece.

At the time of the experiment, the explanation of the unexpected results was that the transverse and longitudinal beams arrived more or less at the same time and therefore the space is not filled by the *aether*. This possibility was disputed later by the corrected calculations, which clearly confirmed that the difference in arrival time of beams is real and valid.

Although this is a serious flaw in the design of the experiment, the main problem was that the beam of light was treated as a one-dimensional entity, adequate for some theoretical calculations. Since one-dimensional entities do not exist in our world, therefore this approach is not applicable for real situations.

In our physical world the source of light does not consist of one singular point, sending only a single, one-dimensional beam. It consists of many of such points, each sending a unique light beam towards the mirrors. These point sources are spread over the whole three-dimensional area of the light source, and therefore the created beams are not synchronized and the distance they travel also slightly differs.

It was wrongly assumed that only one transverse beam of light, reflected from the transverse mirror, will reach the *observer* at point *F*.

In reality, the *observer* could see some other beams, reflected from the transverse mirror, together with the returning longitudinal beam.

This explanation agrees with the results achieved by the *Michelson-Morley* experiment and explains its apparent failure. Unfortunately, the failure to achieve the expected results was at that time interpreted as non-existence of the *aether*, which by itself could not be considered as a valid explanation.

For the experiment's apparently peculiar results, some other explanation had to be found:

'Various suggestions were offered, but, in the light of future investigations at any rate, none of them was so satisfactory or far-reaching as the most revolutionary of all – the principle of relativity.' [1]

In this experiment three basic attributes were analyzed:
- the light and its speed,
- the speed of moving apparatus,
- and the flow of Universal Time.

By consensus it was incorrectly agreed that the culprit must be the time. *Albert Einstein* then included this conclusion in his *special theory of relativity*. He wrote about this experiment:

'If the Michelson–Morley experiment had not brought us into serious embarrassment, no one would have regarded the relativity theory as a (halfway) redemption.'

However, the failure of this experiment cannot be explained by slowing down the rate of flow of *Universal Time*, which is not possible. Furthermore, it proves nothing about the *aether*, which was the intended aim of the experiment.

The results of this experiment do not eliminate the existence of the *aether*, but they also do not prove that it exists. Despite all these shortcomings, the *Michelson–Morley* experiment is still one of the fundamentals of the *special theory of relativity*.

It is obvious that the failure of this experiment cannot be explained by slowing the rate of the *universal time*.

After this experiment, the interest in the behavior of the light had intensified and in 1908, Walther Ritz suggested that the light progresses through the space with constant speed c, relative to its source.

[1] Taken from *'Relativity for All'* by *Herbert Dingle*.

That was refuted in 1913 by *Willem de Sitter*, who based his conclusions on observations of a double-star system. He reasoned that if the speed of light c was only relative to its source, then if observed from different parts of the orbital path, the light from the star would travel away and toward us at different speeds.

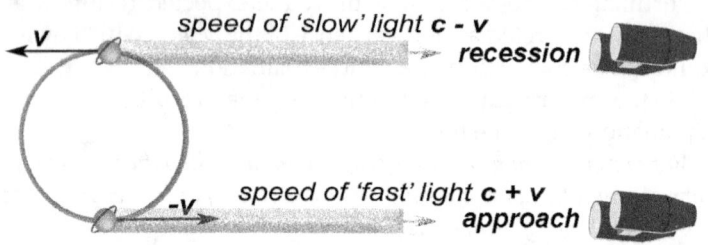

FIGURE A3.8 W. de Sitter - double star system. During approach the revolving star moves toward the observer, and during recession it moves away with speed $\underline{v}$.

If the light emitted by the orbiting star changes its speed at which approaches the *observer* on our planet, then the *observer* would see that '*the "fast" light given off during approach would overtake "slow" light, emitted during a recessional part of the star's orbit.*'[1]

Since this is not the case, *Willem de Sitter's* observations imply that the light must be propagating with its constant speed, regardless of the speed of its source. In other words, he stated that the light propagates in space with a constant speed c, which is independent of the speed of the light source.

This plausible conclusion was incorrectly interpreted as '*nothing can move faster than the light*' and '*no matter how fast the observer travels, the light will be always passing it with its constant speed*'.

Albert Einstein also characterized these observations in his book as: 'VII THE APPARENT INCOMPATIBILITY OF THE LAW OF PROPAGATION OF LIGHT WITH THE PRINCIPLE OF RELATIVITY... *By means of similar considerations based on observations of double stars, the Dutch astronomer De Sitter was also able to show that the velocity of propagation of light cannot depend on the velocity of motion of the body emitting the light. The assumption that this velocity of propagation is dependent on the direction "in space" is in itself improbable.*' [2]

[1] From Wikipedia, W. de Sitter - double star system.
[2] Book 'RELATIVITY THE SPECIAL AND GENERAL THEORY' by Albert Einstein, 1920

The special theory of relativity is with us for almost a hundred years, and it seems that it is infallible. However, in essence its credibility relies heavily on the failure of the *Michelson–Morley* experiment, on the results of *Willem de Sitter*'s observations of a double-star system and finally, on *Lorentz's* calculations.

The fact remains that the vacuum still limits the speed of light and therefore it must have some speed-limiting properties, as for example the water and glass have.

One of the remarkable differences between the sound waves and the light waves is that with increased density of the restrictive medium, sound speeds up, but the light slows down.

	Sound	*Light*
Vacuum	-----	300,000,000
Air	343	300,000,000
Water	1,490	225,000,000
Glass	5,600	200,000,000

The misinterpretation of the results of *Michelson–Morley* experiment led to some wrong, inexact and even erroneous conclusions. Yet, the explanation is obvious:

- *The Michelson–Morley experiment 'failed', because it was wrongly interpreted. In reality, the experiment progressed exactly as it should have.*

- *The light propagates through the space with its constant speed $\underline{c}$, which is valid only in relation to the restrictive medium in which the light propagates.*

- *The restrictive medium found in the universe is the vacuum, which could be filled by the aether. The experiment did not prove or disapprove aether's physical existence.*

- *We do not know the real characteristics of the vacuum, but we know that it restricts the speed of light to its constant value $\underline{c}$. This ability could be attributed most probably to undiscovered atoms of some matter in space, enabling the light to propagate, while at the same time restricting the speed of propagation.*

Appendix 4: Lorentz's Transformation
Reality continues to ruin my life. (Bill Watterson)

Well before Albert Einstein defined his special theory of relativity[1], Dutch physicist H. A. Lorentz was already attracted by the relationship between the light and time. He conducted an abstract experiment, in which he used the light, propagating relatively to a universal reference frame with constant speed c, to calculate a delay in time caused by the *observer's* movement.

In the first part of his abstract experiment, in the universal reference frame, i.e. relatively to the stationary medium in which the light propagates, a stationary *observer* sends a beam of light over distance $\underline{S}_0$ to a distant mirror, and measures the time $\underline{t}_0$ it takes for the beam to return.

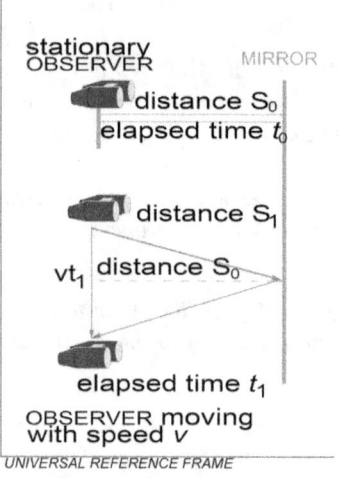

FIGURE A4.1 Lorentz's hypothetical experiment.

In the second part of this experiment, an *observer* moves on a straight line with speed $\underline{v}$, sends a beam of light to the mirror and measures the time $\underline{t}_1$ it takes for the light to return. The light travels the distance $\underline{S}_1$ which is greater than $\underline{S}_0$.

The resulting *Lorentz factor* γ then describes how much longer it takes for the light to reach a moving *observer*, instead of a stationary *observer*.

[1] Book 'RELATIVITY THE SPECIAL AND GENERAL THEORY' by *Albert Einstein*, 1920, Ph.D., translated by *Robert W. Lawson*, M.Sc. University of Sheffield,

$$t_0 = \frac{s_0}{c} \quad t_1 = \frac{s_1}{c} \quad \text{and} \quad s_0^2 + v^2 \cdot t_1^2 = s_1^2 \quad \text{therefore} \quad s_1 = \sqrt{s_0^2 + v^2 t_1^2}$$

$$t_1 = \frac{\sqrt{s_0^2 + v^2 t_1^2}}{c}$$

$$t_1^2 = \frac{s_0^2 + v^2 t_1^2}{c^2}$$

$$t_1^2 \cdot c^2 = s_0^2 + v^2 t_1^2$$

$$t_1^2 \cdot c^2 - v^2 t_1^2 = s_0^2$$

$$t_1^2 (c^2 - v^2) = s_0^2$$

$$t_1^2 = \frac{s_0^2}{(c^2 - v^2)} = \frac{t_0^2 c^2}{c^2 \left(1 - \frac{v^2}{c^2}\right)} = \frac{t_0^2}{1 - \frac{v^2}{c^2}}$$

$$\frac{t_1}{t_0} = \frac{1}{\sqrt{1 - \frac{v^2}{c^2}}}$$

FIGURE A4.2 Calculation of Lorentz factor.

v/c	Lorentz Factor Y	v/c	Lorentz Factor Y
0.5	1.15	0.99999	223
0.8	1.7	0.9999999	2236
0.95	3.2	0.99999999	7071
0.98	5.02	0.999999999	22360
0.99	7.08	1.0	? (infinity)

FIGURE A4.3 Some selected values of Lorentz factor.

Lorentz in his calculations assumed that relative to the universal reference frame, the speed of light is constant. He also assumed that the *observer's* clock, measuring the time delay, is not affected by the *observer's* speed. (This possibility was firstly and mistakenly introduced in the special theory of relativity.)

LORENTZ FACTOR

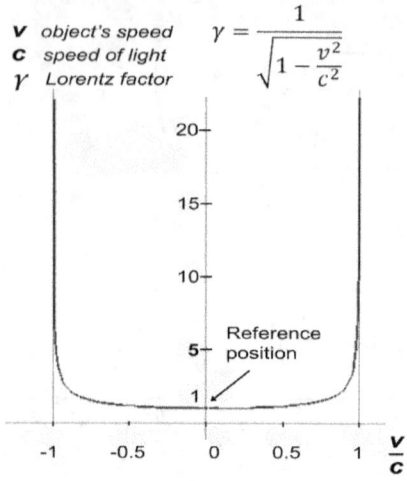

FIGURE A4.4 Graph of Lorentz factor. It uses observers' speed $\underline{v}$ and constant speed of light $\underline{c}$.

It is important to note that all that Lorentz achieved with his calculations was to calculate the time delay. For the light wave, progressing with constant speed, he calculated the time difference between reaching a moving *observer* instead of a stationary *observer*.

It is important to note that in this experiment the flow of time does not change. During travel the same clock with the same rate of time flow is used, as during the stationary part of this experiment. Should the time flow on this clock change, Lorentz's calculations would be meaningless.

If the same clock used for measuring the elapsed time will go slower, there will be no time delay.

In the following years, the first incorrect assumption made by many physicists was to mistake this delay in time for a change in the rate of time flow. They believed that the rate of time flow, measured by the *observer* observing a beam of light, would change due to the *observer's* movement.

The second incorrect assumption was to consider the *Lorentz factor* in only one special case: The *observer* is moving along a straight line, perpendicular to the line connecting the *observer* and the light source from the position closest to the light source. (Illustrated in figure A4.1.)

That excludes any other movements, but in reality, the *observer* could move in whatever direction and from whatever position.

The simplest case to investigate is the *observer's* movement along the line, connecting the *observer* with the light source. The movement could be in both directions, toward and away from the light source. This option traveling on a collision line, directly toward or away from the light source, was missing entirely from *Lorentz's* experiment. This situation is illustrated in figure A4.5.

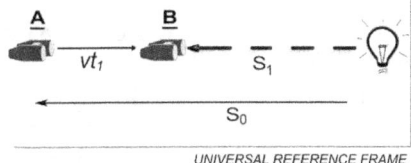

UNIVERSAL REFERENCE FRAME

FIGURE A4.5 *Observer moves from __A__ to __B__ in time __t_1__. The light travels distance __S_1__, and the observer __vt_1__.*

The light emitted by the bulb will reach the stationary *observer* in time t_0 and cover distance S_0. To reach moving *observer*, the light would have to cover distance S_1 in time t_1, while *observer* would travel the distance vt_1

1. Stationary *observer*:
   The light will reach the *observer* at position __A__ in time t_0
   Distance traveled by light will be $\quad S_0 = ct_0$
2. Moving *observer*: Moving with speed __v__ from position __A__ to __B__.
   At __B__ the light will reach *observer* in time t_1
   Distance traveled by the light $\quad S_1 = ct_1 = S_0 - vt_1$
   In time t_1 the *observer* will move $\quad AB = vt_1$

Again, the similar calculations could be used as used by Lorentz:

$$t_0 = \frac{S_0}{c} \qquad t_1 = \frac{S_1}{c} = \frac{S_0 - vt_1}{c}$$

$$t_1 c = S_0 - vt_1$$
$$t_1 c + vt_1 = S_0$$

$$t_1 = \frac{S_0}{c+v}$$

$$\frac{t_1}{t_0} = \frac{\frac{S_0}{c+v}}{\frac{S_0}{c}} = \frac{c}{c+v} = \frac{1}{1+\frac{v}{c}}$$

The figure A4.6 illustrates two graphs, depicting the values of the *Lorentz factor*, and values of its different, modified formula.

Original Lorentz factor: **Extended** Lorentz factor:

$$\gamma = \frac{1}{\sqrt{1-\frac{v^2}{c^2}}} \qquad\qquad \gamma' = \frac{1}{1+\frac{v}{c}}$$

Full line represents the original formula for calculating the *Lorentz factor*. Dotted line represents the extended formula of *Lorentz factor* γ'.

The right side of the dotted line represents a situation where the *observer* moves on a collision course with the light source. In this scenario the *Lorentz factor* γ' will infinitely decrease. That means the light will reach the *observer* in a shorter period of time than in a situation where the *observer* is not moving directly toward the light source.

This dotted right-hand part of the graph is vastly different from other parts and yet, all that makes such a difference is only a very slight change in direction the *observer* travels. The second formula includes not only delay, but also reduction in time interval, needed for the light to reach a moving *observer*.

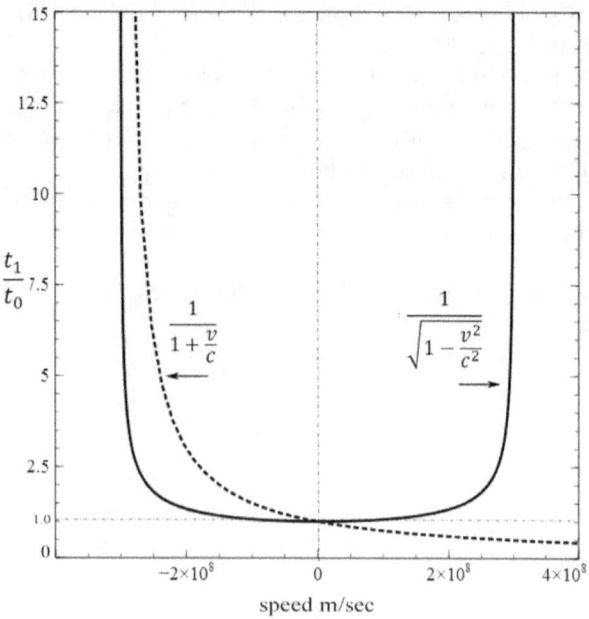

FIGURE A4.6 *Graphs of two versions of Lorentz factor.*

Should we use this extended factor to calculate the rate of time flow, then the time will speed up, which is contrary to what was deduced in the special theory of relativity. This factor is valid only for *observers* moving towards the mirror on a direct line, connecting both the *observer* and the light source.

It is also easy to prove that the *Lorentz factor* will change with the starting position, as illustrated in figure A4 .7.

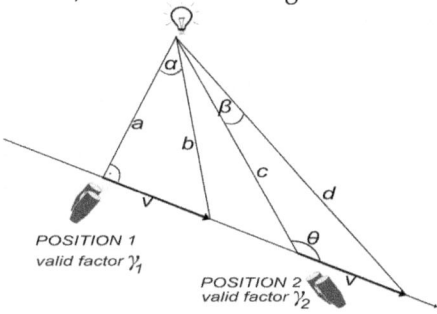

FIGURE A4.7 Observer moves with speed $\underline{v}$ in the same direction. Separate measurements are taken for position 1 and 2.

Starting from position 1, and comparing $\underline{b}$ and $\underline{a}$ will produce the value of originally defined *Lorentz factor* γ. Should the *Lorentz factor* describe movement initiated at any other position on that line, for example position 2, then the ratio of $\underline{d}$ and $\underline{c}$ should be the same as $\underline{b}$ and $\underline{a}$.

Using some properties of a triangle and some trigonometric functions, we could compare these two ratios.

$$\frac{v}{\sin \alpha} = \frac{b}{\sin 90} = b \qquad \frac{v}{\sin \beta} = \frac{d}{\sin \theta}$$

$$\sin(180 - \theta) = \frac{a}{c} = \sin \theta$$

For both factors to be equal: $\dfrac{b}{a} = \dfrac{d}{c}$

$$\frac{\frac{v}{\sin \alpha}}{a} = \frac{\frac{v \sin \theta}{\sin \beta}}{c}$$

$$\frac{1}{a \cdot \sin \alpha} = \frac{\sin \theta}{c \cdot \sin \beta}$$

$$\frac{1}{a \frac{v}{b}} = \frac{\sin \theta}{c \cdot \sin \beta}$$

$$\frac{b}{av} = \frac{\frac{a}{c}}{c \cdot \sin \beta} = \frac{a}{c^2 \sin \beta}$$

$$\frac{b}{a} \neq \frac{av}{c^2 \sin \beta}$$

We can choose any position and evidently the angle β and resulting distance c could have any value, provided $a > \beta$. As a consequence of that, the value of a/b, which proportionally represents *Lorentz factor* Υ, could vary with the position and can have an infinite number of values. The correct formula for the *Lorentz factor* Υ would be also different and would have to include the angles β and θ.

The following example in figure A4.8 illustrates the general case, when the *observer* could move in any direction, not just on the line perpendicular to the line, connecting the *observer* and the light source.

To reach the *observer*, the light has to travel a different distance, and therefore it will reach the *observer* with different delays.

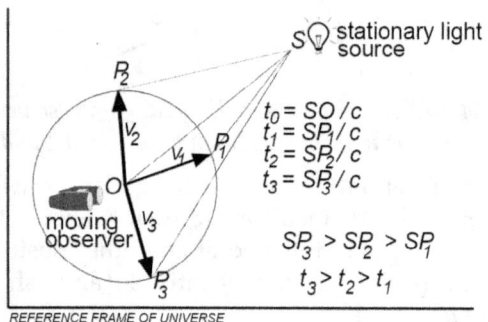

FIGURE A4.8 *The observer could move with the same speed v to any of the positions P_1, P_2 and P_3.*

The observer starts from position O and could move to positions P1, P2 and P3.

The possible distance traveled: $OP_1 = OP_2 = OP_3$

When the *observer* is stationary, the light will travel the distance SO, in time $t_0 = SO/c$

Similarly: $t_1 = SP_1/c$ $t_2 = SP_2/c$ $t_3 = SP_3/c$

Since $SP_1 < SP_2 < SP_3$, the light will travel a shorter time interval, therefore: $t_1 < t_2 < t_3$

The *Lorentz factor* is defined as a ratio of time taken by the light to reach a moving *observer* to time taken to reach a stationary *observer*.

Then, for different directions of travel, and the same *observer's* speed, we would have different values of **Lorentz factor:**

$\Upsilon_1 = (t_1 / t_0)$ $\Upsilon_2 = (t_2 / t_0)$ $\Upsilon_3 = (t_3 / t_0)$
resulting in $\Upsilon_1 < \Upsilon_2 < \Upsilon_3$

These differences are not due to the different *observer's* speed, since $\underline{v}_1 = \underline{v}_2 = \underline{v}_3$, therefore, they would have to be calculated using a different formula for *Lorentz factor* γ.

We have already calculated one such factor γ', for a simplified situation and obviously, the difference is substantial.

It is obvious that the values of the *Lorentz factor* depend not just on the *observer's* speed $\underline{v}$, but also on the position and direction the *observer* is heading. The starting position and direction of movement plays a vital role, and if the *Lorentz factor* should be used in any calculations it has to be included in the formula.

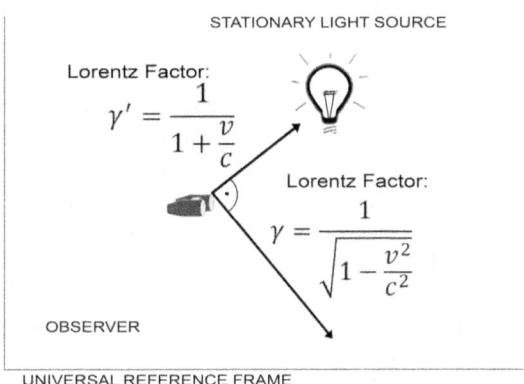

FIGURE A4.9 *Simplified diagram of Lorentz experiment. Observer travels in two different directions and for each situation, different versions of the Lorentz factor exist.*

We can conclude:

- *The Lorentz factor does not represent any changes in the rate of time flow. Furthermore, it is incomplete and has no use in any real-world calculations.*
- *To use the Lorentz factor to define the relativistic mass is erroneous.*

Appendix 5: More on Light

If you only knew the magnificence of the 3, 6 and 9, then you would have the key to the universe. (Nikola Tesla)

The light is the visible part of electromagnetic radiation and is used not just for observing objects, but it is also included in experiments, supporting many theories related to our physical world. For example, the color of the light emitted by a planet is used to estimate the speed of the planet's movement.

For such theories to be correct, the light has to behave the same way in all situations. In chapter 12 and in previous appendix A4 we have already established this is not the case and we introduced and explained the concept of *observed light*.

Although the light has many constant attributes, it is far from being one uniform entity. Even discovered light's constants are not set in stone and their values vary with the nature of light and the medium, in which the light propagates. For example, the *observed light* has its speed influenced by the movements of the *observer* and its color by the movement of its source.

To explain it, we could consider possible combinations of movements of the *observer* and light source, both relative to the surrounding medium.

Firstly, we consider the movement of the light source while the *observer* is stationary, and secondly the movement of the *observer*, while the light source is stationary.

1. Moving light source on direct line with a stationary *observer* (described by *Doppler Effect*):

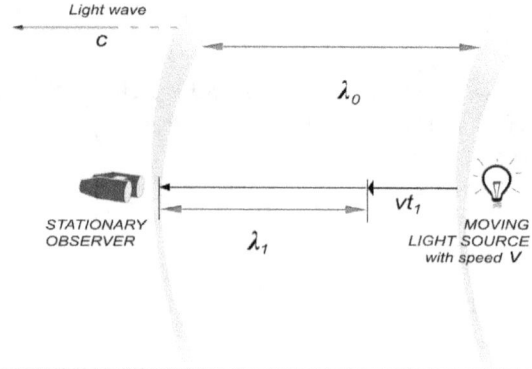

FIGURE A5.1 In a restrictive medium reference frame the light source is moving and the observer is stationary.

In the reference frame of the medium, the light source is moving on a straight line toward the stationary *observer* with the speed v. Since the medium restricts the maximum speed of light to c, the light will reach the stationary *observer* with this constant speed, regardless of the speed v of the source.

The initial, unchanged wavelength of *progressing light* is λ_0, frequency f_0 and speed c. The modified wavelength, reaching the stationary *observer* becomes λ_S and frequency f_S. In time t_1, the light source travels distance vt_1.

For each emitted wave with frequency f_0, the source will travel the distance vt_1

The wavelength then becomes $f_0 = \dfrac{1}{t_1}$ $t_1 = \dfrac{1}{f_0}$

$$\lambda_S = \lambda_0 - vt_1 = \frac{c}{f_0} - v\frac{1}{f_0} = \frac{c-v}{f_0}$$

and frequency $$f_S = \frac{c}{\lambda_S} = \frac{c}{\frac{c-v}{f_0}} = f_0 \left(\frac{c}{c-v}\right)$$

Moving toward the *observer*: $f_S = f_0 \left(\dfrac{c}{c-v}\right)$ ... eq. 1

and away: $f_S = f_0 \left(\dfrac{c}{c+v}\right)$

Obviously, the speed of moving source of light through the restrictive medium affects the real frequency and wavelength of *progressing light*.

2. Moving *observer* on direct line with a stationary light source:

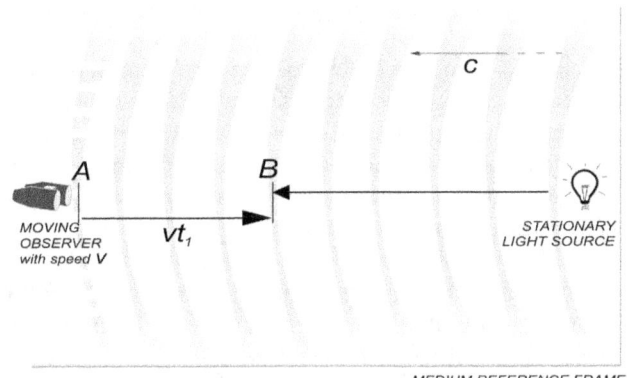

FIGURE A5.2 *In a restrictive medium reference frame, the observer is moving and the light source is stationary.*

The *observer* is moving on a straight line toward the stationary *light source* with the speed v. The medium is stationary and restricts the speed of light to its constant value c.

The wavelength of the *progressing light* is λ, its frequency f_0, speed c. In time t_1 the *observer* would travel the distance vt_1 reaching position B, and the light travels the distance ct_1.

The number of waves the stationary *observer* receives will be: ct_1 / λ, plus the additional number of waves: vt_1 / λ

That will change the *observed frequency* f_{OB} of the light emitted by the light source:

$$f_{OB} = \frac{\frac{ct_1}{\lambda} + \frac{vt_1}{\lambda}}{t_1} = \frac{c+v}{\lambda} = \frac{c+v}{\frac{c}{f_0}}$$

Selecting $t_1 = 1$

$$f_{OB} = f_0 \left(\frac{c+v}{c}\right)$$

Changed frequency f_s due to the movement of the light source is:

$$f_s = f_0 \left(\frac{c}{c \mp v_s}\right)$$

The minus sign holds for the movement toward the *observer* and the plus away from the *observer*.

Changed frequency due to the movement of the *observer* is:

$$f_{OB} = f_0 \left(\frac{c \pm v_{OB}}{c}\right) \quad \ldots \text{eq. 2}$$

The plus sign holds for the movement toward the source and the minus away from the source.

In this scenario, the frequency will change, and the wavelength remains the same[1]. That would affect the wave's speed:

$$c_{OB} = f_{OB} \lambda$$

This speed c_{OB} is the speed of *observed light*, which is the speed of light experienced by a moving *observer*. The speed of *progressing light* remains c, the constant speed of the light in the vacuum.

We could conclude that the speed of an *observer* moving through the restrictive medium affects the frequency and the speed of *observed light*, but not its wavelength.

There is also the third possibility, when the *observer* and the light source are both moving simultaneously. In the following illustrations we simplify the whole situation by assuming that the *observer* and light source moves in two dimensions only, and all movements proceed along a straight line, connecting them both.

[1] See chapter 12 and figure 12.6

2. Moving *observer* and moving light source on direct line:

When both, the source and the *observer* move through the restrictive medium, the combined frequency is

$$f_{COMB} = \left(\frac{c \pm v_{OB}}{c}\right)\left(\frac{c}{c \mp v_s}\right) = \frac{c \pm v_{OB}}{c \mp v_s} \quad \ldots eq.\ 3$$

When moving toward each other, the frequency will be higher, and vice versa.

Scenarios **1** and **2** illustrate two different approaches when considering a propagating wave through some restrictive medium.

In scenario **1**, the wave's frequency is the real frequency, relative to the reference frame of the restrictive medium. Since the speed $\underline{c}$ remains the same, with the change in frequency the wavelength would have to change.

In scenario **2**, the wave's frequency is the observed frequency and differs from the real frequency. This observed frequency is experienced only by a moving *observer*, and each differently moving *observer* would experience a different observed frequency. Since the wavelength remains the same, the observed speed $\underline{c}_{OB}$ of the light would differ from c.

We now have two concepts of *propagating light*:

1. The *propagating light* in a restrictive medium with its constant speed relative to the medium, and with stationary or moving light source.

2. *Observed light* - the same light, being observed by a moving *observer*. The observed speed and observed frequency are both affected by the *observer*'s movements.

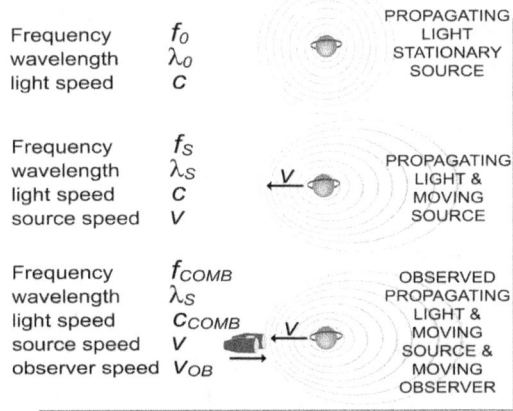

FIGURE A5.3 *Two basic concepts of light.*

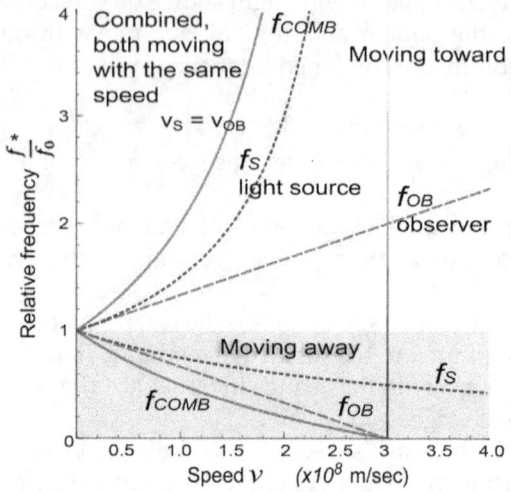

FIGURE A5.4

Change in frequency due to movements of the light source and the observer.

The above graphs represent changes in frequency caused by the movement of the light source and the *observer*. Moving light source is changing the real frequency of the *progressing light* and moving *observer* is changing only the observed frequency.

The dotted line represents the light source moving in relation to the stationary *observer*, and the dashed line represents the *observer*, moving in relation to the stationary light source.

The shadowed area contains graphs, depicting the light source and *observer* moving away from each other. Clear area contains graphs depicting the light source and the *observer* moving toward each other.

Full line is included as an example only. It represents movements of both the light source and the *observer*, both moving with the same speed in relation to the universal reference frame.

The usual situation in our world is when both scenarios are combined. The light source and the *observer* do not have to move on a direct line connecting them, as is illustrated in the following example. Relatively to the restrictive medium, the source of light and the *observer* could move in any direction.

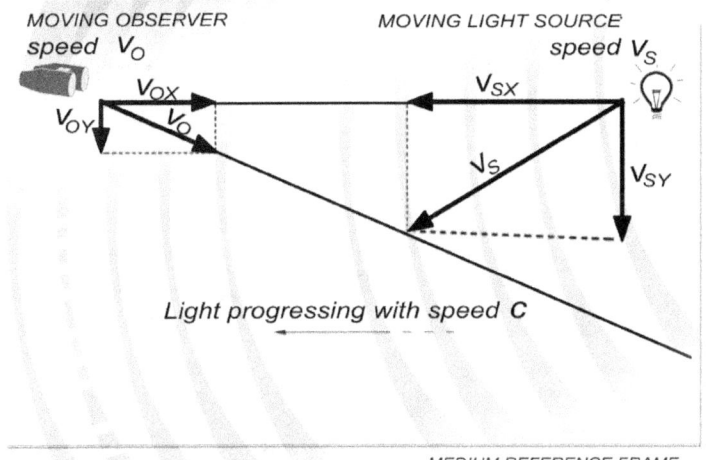

FIGURE A5.5 *Two-dimensional, simplified representation of a typical situation: Observer moves with velocity $\underline{v}_O$ and the source of light with $\underline{v}_S$.*

To obtain the exact *observed frequency* in the above scenario, we would have to know angles $\underline{a}$ & $\underline{\beta}$, and in relation to the medium the speed of both, the light source and the *observer*. This applies to situation, when the source and the *observer* move in two-dimensional plane. Realistically this is seldom the case and we would have to include in calculations the tree-dimensional components of both movements.

This would make the calculations even more difficult and it is obvious that general principle of relativity does not apply to these described scenarios:

Movement of the light source, relative to the stationary *observer*, does not produce the same results as the same movement of the *observer*, relative to the stationary light source.

In the following diagram in figure *A5.6*, the light source is moving away from the stationary *observer* with speed $\underline{v}$, and the light progresses with its constant speed $\underline{c}$.

When the *observer* is moving with the same speed away from the stationary light source, the frequency will differ from the frequency changed by the movement of the light source, moving away from a stationary *observer*. That will produce different *observed* frequency and different observed speed of light.

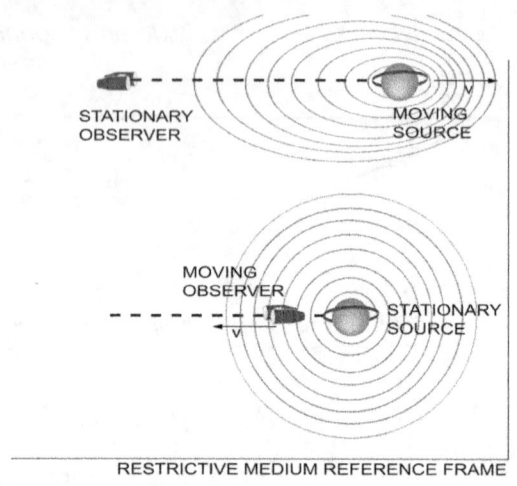

FIGURE A5.6 Effect on the frequency of emitted light by moving light source compared to moving observer.

These scenarios do not apply only to the light, but they also apply to radio waves, etc. That means that any radio signals received from the space could have their original frequency unrecognizably distorted.

To illustrate the complexity of establishing the frequency of observed incoming radiation, we could use the following figure A5.7, depicting the simplified scenario.

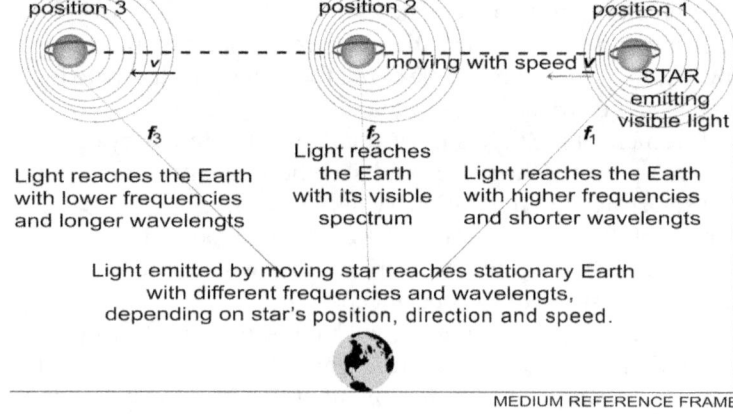

*FIGURE A5.7 The planet is moving from **position 1** to **position 3**. The medium and the Earth are stationary.*

116

In this example the radiation source and the *observer* are only two-dimensional and to simplify it even further, the *observer* is stationary and the movement of the radiation source is in relation to the universal reference frame. We also assume that the unmodified radiation emitted by the star is within the visible spectrum.

The changes in the observed frequency f_1, f_2, f_3 of emitted light from this moving star are caused by the position, direction and speed of the source, measured relatively to the universal reference frame containing the stationary Earth.

Depending on the speed and direction of movement of the star, relative to the medium reference frame and the Earth, the observed radiation, emitted initially by the star as a light, might be reaching the Earth with different frequencies and wavelengths.

In the example above, the following scenarios might apply:

- When the *observer* on the Earth receives a visible frequency of the emitted light from the star, it is an indication that relative to the Earth the star is more or less stationary or it is in a closest position of its trajectory.
- When the *observer* receives only ultraviolet, X-rays, gamma rays or cosmic rays, then the star sending it could be moving toward the Earth.
- When the *observer* receives infrared, microwaves or radio waves, then the star sending it could be moving away from the Earth.

Should we mark the change when the emitted light changes from blue to red color or vice versa, we could establish the moment when the moving star is closest to the Earth.

This example represents a simplified situation that does not have to be always the case. The star due to its speed and direction of travel could also radiate some invisible spectrum of radiation.

It is also obvious that the received frequency will be shifting continuously. Should we monitor in our example a selected frequency of radio waves, sent by a moving planet, initially the receiver will be silent. Due to the motion of the planet, the radio waves will change to higher frequency waves, i.e. Gamma rays, X-rays, light. Later, when the planet is closer to the Earth, and the frequency will reach the tuned frequency of the radio receiver, we will get a *'signal's burst,'*, and then silence again. The actual values of these frequencies depend on the relative speed, direction and mutual position of the planet and the Earth.

It is obvious that from the observed frequencies of received radiation we could calculate only a very vague approximation of stars' speed.

We can conclude:

- *Any theories and calculations involving observed properties of electromagnetic radiation, emitted by some distant planets, are imprecise.*

Appendix 6: The Expanding Universe
All generalizations are false, including this one.
(Mark Twain)

In chapter 16 we have established logical arguments for excluding the possibility of expanding the universe. In this appendix we shall now prove it again, using some simple and only necessary calculations. Firstly, we should examine what exactly Hubble calculated and how.

Hubble's calculations established what is known as the Hubble's law:

$\underline{v} = H_0 d$

v = recessional velocity of galaxy
H_0 = Hubble constant
d = distance to galaxy

$\underline{H_0}$ is *Hubble constant*, approximately equal to 71 km/Mpc [1] (Mpc = million parsec, parsec = 3.26 light years).

What is relevant to our task is that Hubble calculated the speed of moving galaxy $\underline{v_s}$, using light's redshift $\underline{z}$, obtained by comparing the measured wavelength of incoming light:

$$\underline{z} = \frac{\Delta\lambda}{\lambda_0} = \frac{v_s}{c} \qquad\qquad v_s = c\,\frac{\Delta\lambda}{\lambda_0} \qquad\qquad \ldots eq.\ 1$$

$\Delta\lambda$ = measured difference in wavelength (the shift),
λ_0 = initial wavelength,
$\underline{c}$ = constant speed of light in vacuum.

It is immediately obvious that the first hurdle will be establishing the initial, original wavelength of the incoming light.

---

[1] Not used in this book, included only, for better understanding of the whole issue.

The observed galaxy could emit any part of the electromagnetic spectrum, depending on many factors, like position, velocity at which the galaxy progresses through the universe, the direction of travel, etc.

Should we assume that the galaxy is emitting the visible spectrum of the light, and the *observed light* is red, then according to Hubble, the speed of the moving galaxy would be:

$v = 3 \times 10^8 \times (150/550) = 0.82 \times 10^8$ m/sec (= $0.82\underline{c}$).

(Red light wavelength = 700×10^{-9} m,

average white light wavelength = 550×10^{-9} m.)

Should the galaxy initially emit blue light, then the speed of the moving galaxy would be:

$v = 3 \times 10^8 \times (300/700) = 0.43 \times 10^8$ m/sec (= $0.43\underline{c}$).

(Red light wavelength = 700×10^{-9} m,

blue light wavelength = 400×10^{-9} m.)

It is, of course, impossible to know the initial wavelength of the incoming radiation, and therefore this calculation of the speed of the observed galaxy is not reliable.

In the previous appendix we have already expressed doubts about accuracy of any such calculations, involving the observed changes in the frequencies of *observed light*.

This applies to *eq. 1*, which could be also obtained using the *Doppler Effect* calculation, derived in appendix *A2*.

This already derived equation express the change in frequency of light, emitted from a source moving away from the stationary *observer*:
$$f_S = f_0 \left(\frac{c}{c + v_S}\right)$$

Substituting $\underline{f_0} = c/\lambda_0$ and $\underline{f_S} = c/\lambda_S$

$$\frac{c}{\lambda_S} = \frac{c}{\lambda_0} \left(\frac{c}{c + v_S}\right) \quad \frac{\lambda_0}{\lambda_S} = \frac{c}{c + v_S} \quad \lambda_0 v_S = c\lambda_S - c\lambda_0$$

$$v_S = \frac{c(\lambda_S - \lambda_0)}{\lambda_0} = c \frac{\Delta\lambda}{\lambda_0} \quad \text{which is the same as Hubble's } eq.\ 1.$$

The *Doppler Effect* calculations describe only general situations, where the source of light, and/or the *observer* move relatively to each other, along a direct line. However, to comply with the expanding universe scenario, they have to move:
- in two dimension only, i.e. in one plane, and on the line connecting them both with the center of the universe,
- with the same speed and in opposite directions to each other.

To accomplish that, we would have to include in our calculations the *observer's* movements. Then the incoming light becomes the *observed light*, with changed frequency and speed. Hubble just compared the wavelength of light radiated from the galaxy and disregarded the changed frequency and speed, induced by the movements of the *observer*.

Let's assume that by some lucky coincidence, Hubble managed to get readings from a galaxy, aligned with the Earth and with the center of the universe. This situation is illustrated in the figure A6.1.

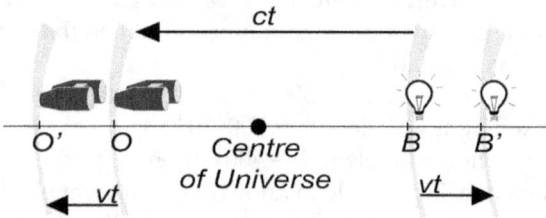

FIGURE A6.1 The desired situation: Observer and the light source are moving with expanding universe along a straight line, connecting them to the center of the universe. Both move with the same speed and in opposite directions.

He would also have to make certain that the galaxy moves away from the center of the universe with the same speed $\underline{v}$, and in the opposite direction that the Earth does. That would cause the *observer* to move from position $\underline{O}$ to $\underline{O}'$, and the light source from position $\underline{B}$ to $\underline{B}'$. In an expanding universe the light emitted by the light source in time $\underline{t}$ would cover a longer distance, increasing by twice the distance $\underline{vt}$. We have already established that the vacuum surrounding both, the *observer* and the light source, is a restrictive medium. That would alter the light emitted from the galaxies as the *Doppler Effect* predicts, provided the movement is along a connecting line. For *observer* moving away from the light source, we have already defined a formula for changed frequency, as was derived in appendix 5, eq. 2:

$$f_{OB} = f_0 \left(\frac{c - v_{OB}}{c} \right)$$

This will cause the *observed light* frequency to decrease, the wavelength will remain unchanged, and the speed of *observed light* will then also decrease.

The *observed light*, although in its nature is not different from the light propagating through the universe, is still different because it is observed, i.e. is experienced by the *observer*. The resulting *observed light* frequency is then a combination of frequency altered by the moving light source f_S and by the moving *observer* f_{OB}.

To calculate the frequency change of the light radiating from a distant galaxy, a combination of both changes to the initial light's frequency has to be considered: $f_{COMB} = f_S f_{OB}$

$$f_{COMB} = f_0 \left(\frac{c}{c+v_{OB}+v_S}\right)\left(\frac{c - v_{OB} - v_S}{c}\right)$$

For the expanding universe to be true, the speed of moving galaxy and the speed of Earth have to be both the same, i.e. $\underline{v}_{OB} = \underline{v}_S$, then $\underline{v}_{EXP} = 2\underline{v}_{OB} = 2\underline{v}_S$ and the following applies:

$$f_{COMB} = f_0 \frac{c - 2v_{OB}}{c + 2v_{OB}} \quad \underline{v}_{OB} = \frac{c}{2}\left(\frac{f_0 - f_{COMB}}{f_0 + f_{COMB}}\right)$$

Hubble calculated the speed $\underline{v}_{OB}$ as $\underline{v}_{OB} = c\frac{\Delta\lambda}{\lambda_0}$

this should then also apply:

$$c\frac{\Delta\lambda}{\lambda_0} \neq \frac{c}{2}\left(\frac{f_0 - f_{COMB}}{f_0 + f_{COMB}}\right)$$

This is obviously not the case, and the speed of *observed light* in the expanding universe is different from what was calculated by comparing the wavelengths of incoming light, as Hubble did.

The calculation involving the moving *observer* is not specific to the expanding universe only. It applies to any two objects emitting and receiving radiation in vacuum. This has many practical implications, for example space communications are affected by changing frequencies.

The figure *A6.2* illustrates the change in frequencies of initially visible *observed light*, caused by a combined retreating movement of the light source and the *observer*.

For the sake of simplicity, the initial frequency f_0 of the light was selected to be in the middle of the visible light's spectrum.

The frequency f_S is caused by the moving source, f_{OB} by the moving *observer*, and f_{COMB} by combination of both.

As depicted in the graph, the light source emits visible light with medium frequency = 5.9×10^{14} m/sec. To see the *observed light* colored red, the *observer's* speed and the speed of the source, should be approx. 0.22×10^8 m/sec ($0.22\ \underline{c}$), and cannot be much lower or higher.

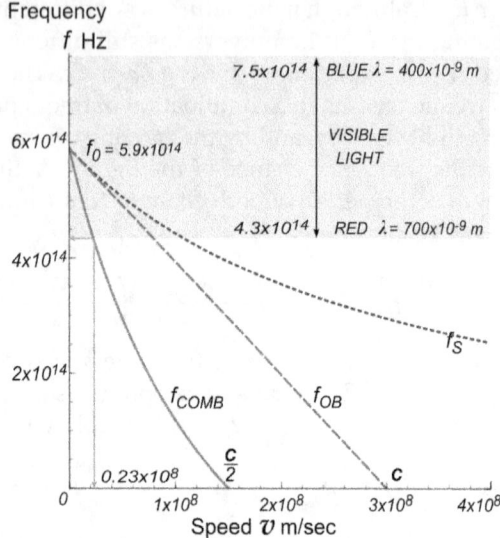

FIGURE A6.2 *A source of visible light and an observer are moving relatively to the vacuum along a straight connecting line, away from each other and with the same speed.*
v representsthe speed of the expanding universe.

The speed below 0.23×10^8 m/sec will not change the light to red color, and any speed above will render the light invisible to human eyes. This seriously limits any chances of seeing a planet radiating the red light and moving along the connecting line.

For a comparison, the speed of the Earth orbiting the Sun is approx. 30,000 m/sec (0.0003×10^8 m/sec). The speed of the expanding universe would have to be then approx. 700x faster than the orbiting Earth.

Obviously, just seeing the red light emitted by some planet does not mean that the planet is traveling at the speed of the expanding universe.

For example, a planet could be emitting the Gamma rays, instead of the light.[1] Due to the modulation of its electromagnetic radiation by the speed of its source and by the speed of the *observer*, the observed progressing radiation could change to the visible light. Depending on the combined speed of the light source and the *observer*, the light could be colored from blue to white and red.

[1] More on this topic in chapter 11.

The figure 6.3, illustrates the case of a moving source of electromagnetic radiation, emitting initially Gamma rays with frequency f0 = 1.5×10^{15} Hz. When the speed of the *observer* and the source reaches 0.5×10^8 m/sec, the observed electromagnetic radiation will become blue light. When the speed reaches 0.85×10^8 m/sec, the color of the *observed light* changes to red.

If the source of radiation and the *observer* both move along the line connecting them to the centre of the universe, then this speed represents the speed of the expanding universe, which in this example will be 0.85×10^8 m/sec.

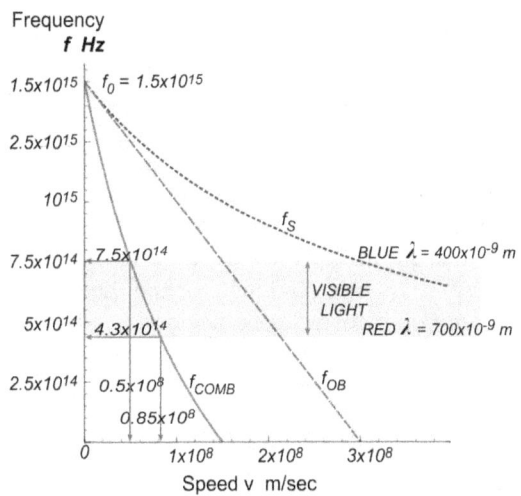

FIGURE A6.3 Source of Gamma rays and an observer are moving relatively to the vacuum along a straight, connecting line and away from each other.

In the previous graph we had the light source initially emitting visible light, setting the speed of the expanding universe to 0.22×10^8 m/sec. The result of Hubble's calculations for that starting frequency equals the speed to 0.82×10^8 m/sec.

For the starting frequency 1.5×10^{15} Hz of Gamma rays, Hubble's calculations set the speed of expansion to 2.14×10^8 m/sec, not 0.22×10^8 m/sec, as depicted by the above graph.

Speed of Expanding Universe m/sec

Starting Frequency f_0	Hubble Calculations	This Book	Difference
5.9×10^{14} Hz	0.82×10^8	0.22×10^8	0.6×10^8
1.5×10^{15} Hz	2.14×10^8	0.85×10^8	1.29×10^8

The difference between these two examples is caused by the selection of the initial frequency of emitted electromagnetic radiation. For the same speed, the nature of the emitted electromagnetic radiation could produce different values for the observed frequency. That represents another discrepancy in the theory of expanding universe.

Since the universe is three dimensional, this shift in frequency could be also attributed to a different combination of the relative movement in the planes, which are different from the plane containing all, the light source, the centre of the universe and the *observer*.

What that all means is that observing a planet or galaxy emitting a red light cannot be used to calculate even its speed in relation to the vacuum. A simple comparison between the wavelengths of incoming light to standard wavelengths of visible light also does not take into account the movements of the *observer*. The concept of *observed light* is not included, which makes any such calculations meaningless.

Combining these findings with the logical arguments, derived already in chapter 16, it is obvious that the theory of expanding universe is based on incorrect assumptions.

We have already proved[1] that our universe is not infinite and logical arguments already derived in chapter 16 point to the non-expanding universe.

We can conclude:

- *Hubble's calculations involve only the movement of the source of light, while the observer is considered being stationary. Equation for the speed of expansion derived by Hubble is different from the equation derived when the observer's movements are included. The inclusion of observer's movements, i.e. the concept of observed light, is necessary for accurate calculations.*

- *It is impossible to make a valid observation of the incoming light and use it for calculating the speed of expansion of the universe. If nothing else, then the simple fact that we do not know the position of the center of the universe makes it unworkable.*

---

[1] More on this topic in chapter 3.

Appendix 7: Changing the Time

The important thing is not to stop questioning. Curiosity has its own reason for existing. (Albert Einstein)

It is still generally believed that the flow of time could be modified, depending on the speed of the *observer*. The faster the *observer* moves, the slower is the rate at which its time flows.

For example, after some years of traveling, an intergalactic traveler returns to the Earth and while his watch displays his current date and time, the watch on the Earth displays much later date and time. He left his young son behind and on his return his son was older than him - a typical fiction topic, of course.

Although we already proved this is not the case, there is still a glimmer of hope for those who believe it: Yes, it is possible to change the rate of time flow. The catch is that it is not the *universal time* we could manipulate, but only the *subjective time* of each individual *observer*.

This could be demonstrated on the following example:

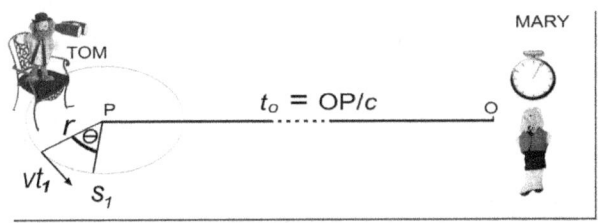

UNIVERSAL REFERENCE FRAME

FIGURE A7.1 *Tom on the chair is moving in a circle. The number of circles is counted by a device, attached to the chair and setting the time rate on Tom's clock to 1 sec for each finished circle.*

It is a hypothetical example and in some ways similar to *Lorentz's* abstract experiment.

Mary and *Tom* are twins, born at the same time. *Mary* is stationary, while *Tom* moves in circles, relatively to her position. *Mary's* clock on the wall measures the *universal time* t_0. *Tom* is experiencing his *subjective time* t_1, set by his electronic watch. At the start, *Tom's* and *Mary's* watches are ticking at the same rate.

Tom starts circling and his electronic device registers the number of circles **_n_** *Tom* travels during 1 sec of elapsed time on *Mary's* watch, and adjusts the time rate on his watch accordingly. Therefore, the speed of *Tom's* watch increases with increasing number of circles *Tom* makes in 1 sec on *Mary's* watch.

For 1 sec of *Mary's* time t_0 *Tom's* time t_1 then becomes:
$$t_1 = nt_0 = n$$
i.e. *Tom's* <u>n</u> seconds becomes 1 sec of *universal time*.

For example, *Tom* makes 3 circles while *Mary's* clock advances by only 1 sec. Then 3 sec of *Tom's subjective time* t_1 becomes 1 sec of *universal time* on *Mary's* clock.

The rate of time flow on *Tom's* clock will be 3x faster than the rate of flow of the *universal time*. As long as *Tom* does not change his speed, this time rate will be his *subjective time* rate.

When *Tom* changes his speed, the number of circles he travels during 1 sec of *universal time* will change.

When *Tom* stops and compares his watch to *Mary's* clock, he finds his clock shows more advanced time and date than *Mary's* clock.

Should *Tom* make less than 1 circle during 1 sec of universal time on *Mary's* clock, his clock will show time and date lagging behind *Mary's* clock.

According to *Tom's* electronic clock, *Tom* is passing *Mary* every 1 sec, so his subjective time is only an illusion and nothing has changed. *Mary's* clock on the wall is still going at the *universal time* rate and *Tom* and *Mary* are still of the same age.

This example is interesting, because it produces the opposite effect for the speeding *observer*: Faster the *observer* moves, faster its subjective time runs. Yet, it is still believed that for such *observers* their time will slow down and not speed up.

The above scenario serves only to visualize in our three-dimensional world our *subjective time*, residing in our mind. The *subjective time* is only a figment of our imagination and does not exist in our universe. It exists only in our mind, which in turn exists in a different world, we already defined as *infinity*.

Since the *subjective time* of an individual *observer* resides in the *observer*'s mind, it could be also changed by many other factors like lifestyle, drugs, boredom, one's mental condition, etc. On the other hand, the *universal time* is an unchangeable part of our world and like other defined constants, it cannot be changed by human intervention.

"Maybe in our next lives ..."
My grandmother Františka

The Predicament Song

We live in dimensions, three and the time,
We measure our success by mountains of gold.
We segregate ourselves into who's strong or frail,
Who's young and beautiful and who's ugly and old.

We became servile, content and weak,
Our freedom is dying by thousands of cuts.
We have no spine to hold our head high,
We have no courage, we have no guts.

We look for the features, so easy to see.
We don't see beauty, hidden in soul.
We value only the obvious traits,
We wrongly judge people, simply as all.

We introduced music, with no melodies,
We introduced art, for those who can't paint.
We regard artists as demigods,
We proclaimed a sinner being a saint.

To sustain our living we fabricate life,
We muster poor creatures through the devil's farm gate.
They never see sunshine, they never feel rain,
What they all live for, is a meal on our plate.

We're heading for our destruction, like a runaway train,
Without reverse and without brakes.
No moral restraints, we don't look back,
Just always forward, whatever it takes.

Our future is written all around us,
As ancient worlds all turned to dust.
We live like culture producing poison,
But we don't believe it, and in lies we trust!

www.ingramcontent.com/pod-product-compliance
Lightning Source LLC
Chambersburg PA
CBHW070647220526
45466CB00001B/330